Algebra Structure Sense Development amongst Diverse Learners

I0131833

This volume emphasizes the role of effective curriculum design, teaching materials, and pedagogy to foster algebra structure sense at different educational levels.

Positing algebra structure sense as fundamental to developing students' broader mathematical maturity and advanced thinking, this text reviews conceptual, historical, cognitive, and semiotic factors, which influence the acquisition of algebra structure sense. It provides empirical evidence to demonstrate the feasibility of linking algebra structure sense to technological tools and promoting it amongst diverse learners. Didactic approaches include the use of adaptive digital environments, gamification, diagnostic and monitoring tools, as well as exercises and algebraic sequences of varied complexity.

Advocating for a focus on both intuitive and formal knowledge, this volume will be of interest to students, scholars, and researchers with an interest in educational research, as well as mathematics education and numeracy.

Teresa Rojano is Professor in the Department of Mathematics Education, Center of Research and Advanced Studies (Cinvestav) of the National Polytechnic Institute, Mexico.

Routledge Research in STEM Education

The *Routledge Research in STEM Education* series is home to cutting-edge, upper-level scholarly studies and edited collections covering STEM education.

Considering science, technology, engineering, and mathematics, texts address a broad range of topics including pedagogy, curriculum, policy, teacher education, and the promotion of diversity within STEM programmes.

Titles offer dynamic interventions into established subjects and innovative studies on emerging topics.

Teaching Early Algebra through Example-Based Problem Solving
Insights from Chinese and U.S. Elementary Classrooms
Meixia Ding

Gender Equity in STEM in Higher Education
International Perspectives on Policy, Institutional Culture, and Individual Choice
Edited by Hyun Kyoung Ro, Frank Fernandez, and Elizabeth Ramon

Mathematics Education for Sustainable Economic Growth and Job Creation
Edited by David Burghes and Jodie Hunter

Queering STEM Culture in US Higher Education
Navigating Experiences of Exclusion in the Academy
Edited by Kelly J. Cross, Stephanie Farrell, and Bryce Hughes

Science and Technology Teacher Education in the Anthropocene
Addressing Challenges in the South and North
Edited by Miranda Rocksén, Elaosi Vhurumuku, Maria Svensson, Emmanuel Mushayikwa, and Audrey Msimanga

For more information about this series, please visit: www.routledge.com/ Routledge-Research-in-STEM-Education/book-series/RRSTEM

Algebra Structure Sense Development amongst Diverse Learners

Theoretical and Empirical Insights to Support In-Person and Remote Learning

Edited by Teresa Rojano

Routledge
Taylor & Francis Group

NEW YORK AND LONDON

First published 2022
by Routledge
605 Third Avenue, New York, NY 10158

and by Routledge
4 Park Square, Milton Park, Abingdon, Oxon, OX14 4RN

*Routledge is an imprint of the Taylor & Francis Group,
an informa business*

Library of Congress Cataloging-in-Publication Data
A catalog record for this title has been requested

ISBN: 9781032055107 (hbk)
ISBN: 9781032055114 (pbk)
ISBN: 9781003197867 (ebk)

DOI: 10.4324/9781003197867

Typeset in Sabon LT Std
by KnowledgeWorks Global Ltd.

Contents

List of Tables

List of Figures

List of Contributors

Eirini Geraniou is an associate professor of mathematics education at the UCL Institute of Education, University College London, London, United Kingdom and a member of the Mathematics Education Group (MEG) at UCL. Her research interests involve the use of digital technologies for the learning and teaching of mathematics, the design and implementation of bridging resources and media for mathematical learning with digital technologies, students' mathematical thinking and computational thinking, students' and teachers' digital competencies. For more information, please visit her UCL IRIS profile.

Carolyn Kieran (Ph.D., McGill University, Montréal) is Professor Emerita of Mathematics Education at Université du Québec à Montréal. Her research focus is the teaching and learning of algebra and algebraic thinking from the primary through to the tertiary levels of schooling. During her career in the Mathematics Department, she spearheaded the creation of the *Algebra in Partnership with Technology in Education* group, a group that she directed for 25 years. The results of her research have been published in over 200 articles, and in book chapters of the *Handbook of research on mathematics teaching and learning* (1992), the *8th International Congress on Mathematical Education: Selected Lectures* (1996)*, The future of the teaching and learning of algebra: The 12th ICMI study* (2004), and the *Second handbook of research on mathematics teaching and learning* (2007). Her contributions to the international mathematics education research community include serving as Chair of the Klein and Freudenthal Awards Committee for the International Commission on Mathematical Instruction (ICMI) (2011–2016), as President of the International Group for the Psychology of Mathematics Education (PME), as invited member of the Mathematics Learning Study committee funded by the National Science Foundation, and as elected member of the NCTM Board of Directors.

Cesar Martínez-Hernández was a young, promising researcher whose life was sadly taken from him by COVID-19 in December 2020. His doctoral research involved the use of digital tools in the learning of school algebra and the role that such tools can play in the co-emergence of technical and theoretical knowledge in algebra. He obtained his Ph.D. in 2013 from the Departamento de Matemática Educativa of CINVESTAV-IPN in Mexico City under the guidance of Professors José Guzman (CINVESTAV-IPN) and Carolyn Kieran (UQAM-Montréal), with a thesis bearing the title: *El desarrollo del conocimiento algebraico de estudiantes en un ambiente CAS con tareas diseñadas desde un enfoque técnico-teórico: Un estudio sobre la simplificación de espresiones racionales.* Within a year or two of completing his doctorate, he obtained an academic post at the University of Colima in Mexico. His 2017–2019 funded project on the development of early algebraic thinking at the primary school level was one that produced such compelling data on the evolution of students' structural thinking about equivalence that it served as the basis for the chapter presented in this volume. But Cesar's interests and research skills did not include just the domain of school algebra; he was equally active in research on the learning and teaching of geometry and calculus. Cesar's gifts for reflection and insight in all of the research that he carried out, and presented both nationally and internationally, will be greatly missed.

Manolis Mavrikis is Professor in Artificial Intelligence and Analytics in Education at UCL Knowledge Lab, Institute of Education, University College London. He holds an MSc with Distinction and Ph.D. from the University of Edinburgh. His research interests and experience are in employing learning analytics to help teachers, schools, and other educators develop an awareness and understanding of the processes involved in learning, and on designing evidence-based intelligent technologies that provide direct feedback to learners. He is currently Director of the MA in Education and Technology and one of the editors for the *British Journal of Educational Technology.*

Valentina Muñoz-Porras is an outreach mathematician and a software engineer. She earned a master's degree in Computer Science from UNAM and a Ph.D. in Mathematics Education from Cinvestav. She specializes in using technologies to design and develop digital activities that allow students to explore, discover, and create meanings to understand mathematics better.

Currently, Valentina works at the Mathematics Research Center (CIMAT) in Guanajuato, Mexico, where she runs professional development workshops and organizes outreach activities. She also teaches mathematics at high school and undergraduate levels. She is the co-author of a mathematics high school textbook and author of multiple interactive contents for learning mathematics.

Santiago Palmas is a researcher in the field of technology and didactics. He has worked in education with children, youth, and adult with little or no schooling. Since 2017, he has been a researcher of the

Cultural Studies Department, Metropolitan Autonomous University – Campus Lerma, where he was temporary Head of the Department in 2020–2021. Co-author of the *Mexican National Textbook on Mathematics*, Santiago has collaborated with the National Secretary of Public Education as a regular consultant for primary mathematics education and adult education. He is member of the National Research System, the Mexican Mathematical Society, the National Council for Educative Research, and some international organizations such as the Comparative and International Education Society and the British Society for Research into Learning Mathematics. In his latest research, he has focused on the socio-political role of mathematics education and the role of mathematics education in the capitalist world.

Luis Puig, Professor Emeritus of Didactics of Mathematics at the University of Valencia, was born in 1948 in Valencia, Spain. He is author of four books, *Problemas aritméticos escolares* (with Fernando Cerdán), *Semiótica y matemáticas, Elementos de resolución de problemas,* and *Educational algebra. A theoretical and empirical approach* (with Eugenio Filloy y Teresa Rojano), and co-author of some 15 books of classroom materials linked to curriculum development projects. He has published articles and given lectures at national and international conferences. He was co-editor of two series of books on mathematics education (*Matemáticas, cultura y aprendizaje, Mathema*) and several series on cultural studies (*Eutopías, Eutopías Mayor, Biblioteca Otras Eutopías, La huella sonora*), and of the journal *Enseñanza de las Ciencias.* He was a member of the *Commission Internationale pour l'Étude et l'Amélioration de l'Enseignement des Mathématiques,* the Internacional Committee of the *International Group for the Psychology of Mathematics Education,* founder President of the *Societat d'Educació Matemàtica de la Comunitat Valenciana "Al-Khwārizmī",* and he is a member of the Executive Committee of the *International Group on the Relations between the History and Pedagogy of Mathematics.* His main areas of work have been curriculum development, heuristics, algebraic problem solving, and the history of algebra.

Teresa Rojano has conducted research in the didactics of algebra with a special focus on the transition from arithmetic to algebraic thought. She has conducted collaborative research with the Bristol University and the London University on school algebra and mathematical modelling within technology environments. Since 1975, she has been based at the Centre of Research and Advanced Studies (Cinvestav) of the National Polytechnic Institute in Mexico, where she was appointed Professor of Mathematics Education in 1985. She was vice-president of the International Group for the Psychology of Mathematics Education (1995–1997); member of the Program Committee of the International Congress of Mathematics Education (2001–2004); leader (with Luis Puig) of the History Group for the Algebra Study of the International Commission for Mathematical Instruction; and director of the Mexican project *Incorporation of New Technologies to the School Culture*. From 1997 to 2003, she was head of the Department of Mathematics Education – Cinvestav and was advisor of the Ministry of Education (SEP) in Mexico from 2003 to 2006. Since 2008, she is a member of the Advisory Board of the James J. Kaput Center for Research and Innovation in Mathematics Education (UMASSD). In May 2009, she was invited to be a member of the International Advisory Board of the *Journal for Research in Mathematics Education* (NCTM). Since 2013, she is member of the Editorial Board of the journal *Educational Studies in Mathematics*, published by Springer; in 2018, she was invited to be a member of the Editorial Board of the journal *Mathematical Thinking and Learning*, published by Routledge. In 2012, Teresa Rojano was appointed Emeritus Professor at Cinvestav.

Armando Solares-Rojas is a researcher at the Department of Educational Mathematics at Cinvestav. His two current lines of research are mathematical modelling and the study of mathematical knowledge in contexts of cultural diversity. In these lines he addresses, respectively, mathematical modelling processes of scientific and social phenomena; and the mathematical knowledge mobilized in activities of diverse cultural groups, from a semiotic and historical-cultural point of view. He has been a visiting professor at various universities in Mexico, Spain, and Canada. He has participated in teacher training projects, national textbooks elaboration, and national school study programmes advising in Mexico, as well as in various Mexican and international research projects. Currently, he is the director of a research project funded by the Engineering and Physical Sciences Research Council (EPSRC) – United Kingdom Research and Innovation (UKRI) (Grant Ref: EP/T003545/1) on mathematical modeling of ecological problems that take place in communities in situations of environmental crisis in Latin America.

Ulises Xolocotzin is a researcher in the Mathematics Education Department at Cinvestav, Mexico. Before this, he was a lecturer at UNAM and a postdoctoral researcher at the University of Bristol and Cinvestav. He completed his studies in Psychology at UNAM, a Master's degree in Educational Psychology at UNAM, and a Ph.D. in the Learning Sciences Research Institute at the University of Nottingham. His research interests include mathematical cognition, affective factors in mathematics learning, and digital technologies in mathematics education.

Foreword

Algebra Structure Sense in Historical Context

David Kirshner

In the foreword, I seek to contextualize this volume's interest in algebra structure sense historically within the broader struggle within mathematics education to make students' engagement with mathematical representations meaningful. The most obvious and persistent sign students engage with algebra without a sense of meaning are "mal-rules" (Sleeman, 1986) that generate not random errors, but systematic overgeneralizations of correct rules (Table 0.1). Such error patterns – widely documented over decades (e.g., Booth, 1984; Bundy & Welham, 1981; Davis, 1979; Payne & Squibb, 1990) – suggest a kind of mindless engagement with the symbol system; in Thompson's (1989) memorable phrase, students are "pushing symbols without engaging their brains" (p. 138).

One might find comfort in the fact that at least some students do overcome mal-rules, especially if one holds to the "iterative view" that skills and concepts are deeply intertwined: "the causal relations [between

Table 0.1 Mal-rules and Correct Rules

Mal-rules	Correct Rules
$(a+b)^c = a^c + b^c$	$(a+b)c = ac + bc$
$\sqrt[c]{a+b} = \sqrt[c]{a} + \sqrt[c]{b}$	$\sqrt[c]{ab} = \sqrt[c]{a}\,\sqrt[c]{b}$
$a^{m+n} = a^m + a^n$	$a(m+n) = am + an$
$\dfrac{a}{b+c} = \dfrac{a}{b} + \dfrac{a}{c}$	$\dfrac{b+c}{a} = \dfrac{b}{a} + \dfrac{c}{a}$
$\dfrac{a+x}{b+x} = \dfrac{a}{b}$	$\dfrac{ax}{bx} = \dfrac{a}{b}$

procedural and conceptual knowledge] are said to be bi-directional, with increases in conceptual knowledge leading to subsequent increases in procedural knowledge and vice versa" (Rittle-Johnson & Schneider, 2015, p. 1126). But we should not be too sanguine about even the "success stories" of secondary school mathematics.

I had the privilege of conducting an evaluation study of an innovative Calculus 2 curriculum at an elite private university in the central U.S. The mean math-SAT score for these students was an astonishing 717 (92nd percentile). Yet their Calculus 2 professors complained bitterly that students lacked the sophistication of being able to read into the structure of rational algebraic expressions to be able to anticipate the behaviour of the functions.

My subsequent investigation found a near total lack of structure sense (Kirshner & Chance, 2005). When asked to find solutions for x of the following pair of equations, fewer than 10% of these students were able see the second equation as a special case of the first.

$$\left(2x + \frac{1}{2}\right)^2 - 2x + 3 = 21\frac{1}{4}, \text{ and}$$

$$\left(2\sin x + \frac{1}{2}\right)^2 - 2\sin x + 3 = 21\frac{1}{4}$$

When given a choice of meanings for the symbol "$f(x)$", all but four out of 122 students included "function of x" as a correct interpretation. Persistent mal-rules in symbol manipulation may be the most obvious indicator of a failure of meaning in algebra but the absence of algebra structure sense runs much deeper than that.

Curriculum Reform Efforts

Though not always discussed in terms of algebra structure sense, the superficiality of learning has long been a concern of mathematics educators. We briefly review two of the most concerted curriculum reform efforts of the past century designed to ameliorate this problem.

One analysis of the problem of superficiality holds that students are not approaching the study of algebra with sufficient formal and logical rigor. The *New Math* era of the 1960s through mid-1970s was oriented by "the concepts of set, relation, and function and by judicious use of broadly applicable mathematical processes like deductive reasoning and the search for patterns" (Fey & Graeber, 2003, p. 524). Teachers were taught abstract algebra (e.g., Haag, 1961), and textbooks of the day included basic proofs of elementary theorems that undergird rules of elementary algebra (e.g., Dolciani & Wooton, 1975). However, the curriculum did not perform as intended, coming to be widely regarded as "excessively formal, deductively structured, and theoretical.... fail[ing] to meet the needs for basic mathematical literacy of average and low ability students" (NACOME, 1975, p. ix).

The lessons of the *New Math* were not lost on mathematics educators. The next major movement, initiated by NCTM's (1989) *Curriculum and Evaluation Standards for School Mathematics*, located meaningfulness for mathematics in contextual settings. As Kaput (1995) put it in an influential paper:

> Acts of generalization and gradual formalization of the constructed generality must precede work with formalisms – otherwise the formalisms have no source in student experience. The current whole-sale failure of school algebra has shown the inadequacy of attempts to tie the formalisms to students' experience *after* they have been introduced. It seems that, "once meaningless, always meaningless."
>
> (pp. 74–75)

The source of this "gradual formalization" as stemming for "contextual settings" is made explicit in NCTM's Algebra Working Group's (1998) Framework:

> The "Framework" proposes a way to develop algebraic reasoning by exploring a variety of contextual settings that are connected by organizing themes. By serving as organizers, themes help students recognize important ideas and make connections. Contextual settings are the ground on which these themes play out. They provide the substance from which and about which to reason.
>
> (p. 164)

Although it is understandable that one might look to contextual settings to ground mathematical meaning, this agenda has never been reconciled with the abstract, decontextualized aspect of algebra. As Bell (1936) put it, "the letters are mere *undefined marks* or 'elements' about which certain postulates are made.... The very point of elementary algebra is simply that it *is* abstract, that is, devoid of any meaning beyond the formal consequences of the postulates laid down for the marks" (p. 144).

I want to argue that these two major movements in mathematics education miss the mark in their analysis of the problems of algebra learning in related ways. Both are correct that superficiality of learning stems from the traditional focus of instruction on repetitive practice of procedural routines. But neither grasps the nature of the deficit created, nor the character of competencies needed to overcome it.

For the *New Math*, the perceived deficit is a lack of logical engagement. If students simply come to think the right thoughts, guided by logic, all will be well. For *Standards* advocates it is not logical thought that is missing but contextual thought. If students can come to focus on the meanings inherent to the contexts studied they will be able to comprehend the abstractions that underlie the symbol system. In both cases, the symbolic forms through which algebraic expressions are represented are neutral carriers of meaning, the challenge being to get to those meanings through one means or another.

While embracing the importance of both logic and context, algebra structure sense is based on the further intuition that symbolic forms are not neutral carriers of meaning; form and meaning are irreconcilably intertwined. So it is that Rojano (this volume) points us towards the "internal order of algebraic *texts* [that] can lead to a better understanding of the difficulties faced by learners and to further opportunities for proposing new teaching approaches that take the inextricable subject-structure relationship into account" (Chapter 1). It is this semiotic dimension, so ably pursued in the chapters of this book, that makes algebra structure sense the new frontier for understanding and developing valued competencies in algebra.

References

Bell, E. T. (1936). The meaning of mathematics. In W. D. Reeve (Ed.), *The place of mathematics in modern education* (pp. 136–181). New York: Eleventh Yearbook of the National Council of Teachers of Mathematics. Bureau of Publications, Teachers College, Columbia University.Bureau of Publications, Teachers College, Columbia University.

Booth, L. R. (1984). *Algebra: Children's strategies and errors.* Windsor, Berkshire, UK: NFER-NELSON Publishing Company.

Booth, L. R. (1989). A question of structure. In C. Kieran, & S. Wagner (Eds.), *Research agenda for mathematics education: Research issues in the learning and teaching of algebra* (pp. 57–59). Reston, VA: National Council of Teachers of Mathematics; Hillsdale, NJ: Lawrence Erlbaum Associates. National Council of Teachers of Mathematics; Hillsdale, NJ: Lawrence Erlbaum Associates.

Bundy, A., & Welham, B. (1981). Using meta-level inference for selective application of multiple rewrite rule sets in algebraic manipulation. *Artificial Intelligence, 16,* 189–212.

Davis, R. B. (1979, April). *Error analysis in high school mathematics, conceived as information-processing pathology.* Paper presented at the annual meeting of the American Educational Research Association, San Francisco. (ERIC Document Reproduction Service No. ED171551).

Dolciani, M. P., & Wooton, W. (1975). *Modern algebra: Structure and method (Module 1),* (revised edition), Markham, Ontario: Houghton Mifflin Canada Limited.

Fey, J. T., & Graeber, A. (2003). From the new math to the agenda for action. In G. Stanic, & J. Kilpatrick (Eds.), *A history of school mathematics* (Vol. 1, pp. 521–558). Reston, VA: National Council of Teachers of Mathematics.

Haag, V. H. (1961). Structure of elementary algebra (revised edition). In *Studies in mathematics* (Vol. 3). New Haven, CT: School Mathematics Study Group. Yale University.

Kaput, J. J. (1995). A research base supporting long term algebra reform. In D. T. Owens, M. K. Reed, & G. M. Millsaps (Eds.), *Proceedings of the Seventeenth Annual Meeting of the North American Chapter of the International Group for the Psychology of Mathematics Education* (Vol. 1, pp. 71–94). Columbus, OH: ERIC Clearinghouse for Science, Mathematics, and Environmental Education.

Kirshner, D., & Chance, B. (2005). Measuring algebraic sophistication: Instrumentation and results. In S. Wilson (Ed.), *Proceedings of the Twenty-Seventh Annual Meeting of the International Group for the Psychology of Mathematics Education*. Roanoke: Virginia: North American Chapter.

National Advisory Committee on Mathematics Education (1975). *Overview and analysis of school mathematics grades k-12*. Washington, DC: Conference Board of the Mathematical Sciences.

National Council of Teachers of Mathematics. (1989). Curriculum and evaluation standards for school mathematics. Reston, VA: Author.

National Council of Teachers of Mathematics, Algebra Working Group. (1998). A framework for constructing a vision of algebra: A discussion document. In *National Research Council, the nature and role of algebra in the k-14 curriculum: Proceedings of a national symposium*, May 27 and 28, 1997 (Appendix E, pp. 145–190). Washington, DC: National Academy Press.

Payne, S. J., & Squibb, H. R. (1990). Algebra mal-rules and cognitive accounts of error. *Cognitive Science, 14*, 445–481.

Rittle-Johnson, B., & Schneider, M. (2015). Developing conceptual and procedural knowledge of mathematics. In R. C. Kadosh, & A. Dowker (Eds.), *The Oxford handbook of numerical cognition* (pp. 1118–1134). New York: Oxford University Press.

Sleeman, D. (1986). Introductory algebra: A case study of student misconceptions. *Journal of Mathematical Behavior, 5*(1), 25–52.

Thompson, P. W. (1989). Artificial intelligence, advanced technology, and learning and teaching algebra. In C. Kieran, & S. Wagner (Eds.), *Research agenda for mathematics education: Research issues in the learning and teaching of algebra* (pp. 135–161). Reston, VA: National Council of Teachers of Mathematics; Hillsdale, NJ: Lawrence Erlbaum Associates. National Council of Teachers of Mathematics; Hillsdale, NJ: Lawrence Erlbaum Associates.

Preface

Algebra structure sense (ASS) is the ability to become aware of structural properties of algebraic objects. Developing this capability is essential to gaining access to advanced algebraic thinking – and mathematical thinking in general, as well as to developing the competencies needed to use symbolic algebra in modelling and solving problems within a variety of contexts. In this book, we establish the need to delve deeper into the analysis of the structural features of algebraic objects, and into the factors that influence algebra language users' awareness of such features.

Conceptual aspects of the notion of ASS are reviewed, and the feasibility of having user populations with different mathematical background develop their ASS by working in appropriate learning environments is established. The plausibility of this approach is supported by the findings of studies that use learning environments in which a central role is played by both the dynamic interconnection of different representations of algebraic objects, and a didactic design organized by levels of syntactic complexity of algebraic tasks, and with digital components of adaptivity and gamification. Unlike studies that suggest ways of verifying whether a person possesses (or not) ASS, here the authors discuss the feasibility of "teaching" or rather of promoting development of ASS among diverse groups of subjects. Thus, the content of Chapters 4 and 5 is based on outcomes from a project entitled *Developing algebra structure sense in a digital interactive environment with an adaptive system*.[1] In this project, the idea of fostering development of different levels of algebraic competencies in heterogeneous groups of students is explored. To this end, a learning environment that provides feedback to users when they engage in transformative algebra tasks has been developed and tested out. The technological and didactic characteristics of the environment are analysed (Chapter 4), as are the results of the experimental work carried out with subjects from different school levels and contexts (Chapter 5). In addition, and following the same line on technological aspects, in Chapter 6, experience is shared from a study on identification of the structure of patterns and its relationship with its numerical and algebraic expressions. The study was carried out using the *eXpresser microworld,*[2] developed by a research team at UCL Knowledge Lab in the United Kingdom.

Besides the content related to the application of learning interactive environments, conceptual and theoretical aspects concerning structure sense are discussed from different perspectives, such as the standpoint of the historical development of algebra (Chapter 2); numerical thinking (Chapter 3), the implications that the development of structure sense, as well as the theoretical reflections on the theme, have on the field of learning and teaching algebra and mathematics in general (Chapter 7); the role of transformational algebra and the connection between the internal structure of the mathematical object and the subject that sees and transforms it (Chapter 8). With this, we intend to contribute to broaden the notion of ASS present in the current specialized literature.

Teresa Rojano

Notes

1 *Developing algebra structure sense in a digital interactive environment with an adaptive feedback system* is a Frontiers of Science Project, funded by the Council of Science and Technology in Mexico (Conacyt Ref. 2016-01-2347). The main aim of the project is to prove the feasibility to foster the development of high algebraic competencies in heterogeneous groups of students, using a technology environment that promotes autonomous learning (https://www.teresarojano.com).

2 The key idea of the *eXpresser microworld* is that students first *identify* the structure of a pattern of squares presented dynamically, next *construct* the pattern, and finally, *express* a general rule for the number of tiles in a general pattern. Thus, there is a tight coupling between building the pattern, and being able to describe how it is built – between the "algebra" and the objects the algebra aims at expressing. In this book chapter we present the version of the system with which students interacted during the studies (http://www.migen.org).

Acknowledgements

I would like to thank: The National Council of Science and Technology (CONACYT) in Mexico for the funding to carry out the 3-year Frontiers of Science Project *Developing algebra structure sense in a digital interactive environment with an adaptive feedback system*; colleagues of the research team, Valentina Muñoz, Santiago Palmas, Armando Solares, Ulises Xolocotzin, and Manolis Mavrikis; Lupita Guevara and Tania Villanueva for their support in different aspects of the project development; the participants in the experimental stage of the project; as well as the contributors of Chapters 2, 3, and 6 for their insights, which enriched the content of this book.

1 Algebra Structure Sense

Conceptual Approaches and Elements for Its Development

Teresa Rojano

In order to advance the conceptualization of algebra structure sense and to conceive didactic approaches that promote its development, we posit that there is a need to deepen the knowledge of the internal structure and order of algebraic objects, as well as the elements that come into play in the processes related to awareness of such structure.

Introduction

Research carried out in the 1960s by Vadim Krutetskii reports several cases of gifted children, one of which is of Sonya L., who was observed between the ages of 8 and 10 and demonstrated (among other skills) a great ability to understand and use the generality of a simple rule. In one of the experiments, Sonya was presented with the rule $5a + 5b = 5(a+b)$ and was then asked to express in a simpler form the expression $4m^2(2p-q) - 2m(q-2p) - 2m(q-2p)$. Her immediate verbal and written response was: 'In two of the parentheses we only need to change the signs' and she wrote $2m(2p-q)(2m+2)$ (Krutetskii, 1976).

Keeping a respectful distance from the analysis that Krutetskii may have made of the above episode from a psychological perspective and in the context of his research, which was aimed at isolating and characterizing a set of mathematical abilities, it can be said that Sonya's immediate response speaks of the presence of a structure sense with respect to algebraic expressions. A concrete expression of the definition of algebra structure sense (ASS) formulated by Hoch (2007) is perfectly captured in the above episode with Sonya L. According to this author, subjects show that they have ASS if they are able to: (a) recognize a structure in its simplest form; (b) treat a compound term as a single entity and, through appropriate substitution, recognize a familiar structure within a more complex form; (c) choose appropriate transformations to make effective use of the structure. In Sonya's performance, phases (a)–(c) are observed in a single step; by the way, Krutetskii calls this ability *abbreviation of thought*.

DOI: 10.4324/9781003197867-1

The use of the operational versions of 'structure sense' as above, which focus on subject actions, also necessarily involves elements rooted in the form and essence of the object (of knowledge), in the case of symbolic algebra the objects are symbols, expressions, and equalities between algebraic expressions. In the latter respect, some authors refer to intrinsic properties of the object, such as grammar (Esty, 1992), the superficial, and deep structures of expressions (Kirshner, 1989) or to the hierarchy, form, and internal order of expressions (Sfard & Linchevski, 1994). While other authors include elements that belong to the subject's relationship with the algebraic object, and such is the case of Kirshner and Awtry (2004) who, in order to give plausible explanations for certain frequent mistakes that students make in algebra, resort to the attribute of the visual salience of an algebraic expression, an attribute that depends both on the properties of the expression and on 'whomever reads it'. All of the above shows a set of perspectives, from which one can delve into the notion of ASS, understood broadly as the ability to be aware of the structural properties of an algebraic object. In this chapter, aspects of this notion that have been studied from some of these perspectives are addressed, and elements for an ASS didactics are identified.

The Objects of Symbolic Algebra

Symbolic algebra is a system of signs whose syntactic rules are conventional; they are arbitrary and not natural; however, the system and its rules, as we know them today, are the result of a long and intricate history. With the caveat that in the next chapter L. Puig takes on a more extensive discussion of this topic, in order to provide an idea of the history behind the language of algebra, it is worthwhile to reference the three stages of the evolution of algebraic symbolism stated by G.H.F. Nesselmann in his 1842 work (Cajori, 1928–1929): the rhetorical algebra stage, where the solution to a problem is written as a prose argument; the stage of syncopated algebra, where stenographic abbreviations for quantities, relations, and operations that occur with greater recurrence are used; and the symbolic algebra stage, where solutions to problems are expressed in a synthetic language, composed of symbols that have no apparent relationship with what they represent (Eves, 1983, p. 126). The abacus texts, which include Fibonacci's *Liber Abbacci* (Sigler (tr.), 2002), are an example of rhetorical algebra. Whereas the Diofanthine algebra, contained in the book *La Arithmetica*, (Heath 1910), and Jordanus de Nemore's *De Numeris Datis* (Hughes, 1981) are two variants of the syncopated stage. Finally, according to various mathematics historians, the appearance of *The Analytical Art* by François Viéte in the 16th century (Witmer (tr.), 1983) marks the birth of symbolic algebra. The following examples from each of these stages show how the relationship between representations and what they represent changes from one stage to another, becoming less explicit as these representations evolve into more symbolic, more abstract forms. However, even in the Vietic syntax, in which the expressions are formed under Zetetic rules[1], the geometric lineage of the expressions can be appreciated.

> Example 1: A problem statement and part of its solution expressed in rhetoric algebra (Problem 1 from the abacus book *Trattato di Fioretti* by M.A. Mazzinghi (1350, Italy) (G. Arrighi, ed., 1967). Pisa, Italy: Domus Galileana) and its translation into current algebraic symbolism (Rojano, 1985, unpublished PhD dissertation, Cinvestav – IPN, Mexico).

Original version in old Italian

Problem 1

Fa' di 19, 3 parti nella proportionalità chontinua che, multiplichato la prima chontro alla altre 2 e lla sechonda parte multiplicato all'alltre 2 e lla terza parte multiplichante all'altre 2, e quelle 3 somme agunte insieme faccino 228. Adimandasi qualj sono le dette parti.

> *Translation into modern symbolism*
>
> Find three numbers, x, y, z such that:

$$x + y + z = 19$$

and that

$$x/y = y/z$$

$$x(y+z) + y(x+z) + z(x+y) = 228$$

Part of the solution in original old Italian version
Conciosiachosachè Maestro Antonio sottile scriva e' chasi e' qualj non ànno asolutione alle quantità numeralj, nientedimeno io mi sforçerò di porre chasi e' qualj aranno asolutione a quantità che facilmente si potranno provare, cioè che con facilità acciò che lo 'nteletto possa essere chapacie di tutto.
    Ora al nostro chaso, qui è da sapere che se si multiplicha 19 nel doppio de la seconda parte sarà 228, overo si se multiplicha lo droppio del 19, cioè 38, nella seconda parte farà 228. E che questo sia vero il voglio chiarire.
    Noi abbiamo proposto che si divida 19 in 3 parti nella proportionalità chontinua che, multiplichata la prima per l'altre 2 e la seconda per l'altre 2 e l'altra, cioè la terça, per l'altre 2 et agunte le dette multiplichatione, insieme faccino 228.

Part of the solution translated into modern symbolism
Solution:
You have to know that

$$(19)(2y) = 228$$

Which is equal to

$$2(19)y = 228$$

$$38y = 228$$

This must be explained.

Example 2: A problem statement and part of its solution expressed in syncopated algebra (Translation of Proposition 2 from Book One of *De Numeris Datis* by Jordanus de Nemore. Acritical edition. B. Hughes (ed./tr.) (1981). University of California Press).

Proposition 1.2

If a given number is separated into as many parts as desired, whose successive differences are known, then each of the parts can be found. Given is the number a which is divided into w, x, y, and z the least of the parts.

Since the successive differences of all of these are given, each difference can be expressed in terms of the difference of each number with z. Therefore, let f be the difference of w and z, and the sum of g and h be the sum of the differences of x and z with y and z. Now because z makes each of those equal to each of these, it is obvious that thrice z with the sum of f, g, and h equals those three. Therefore, four times z with the sum of f, g, and h equals a.

Example 3: An algebraic expression written with the Zetetic's rules (*The Analytic Art* by François Viéte (1540–1603). Witmer, T.R. (tr.) (1983). Kent, Ohio: The Kent State University Press). Below, the same expression is written with modern algebraic notation.

$$\frac{A\ CUBUS - B\ SOLIDO\ 3}{C\ in\ E\ CUADRATIVA}$$

$$\frac{X^3 - 3b}{cY^2}$$

The aforementioned stages did not follow a chronological order, and, in fact, it is thought that rhetorical algebra was predominant for a long time in the work of Western European algebraists (Eves, 1983, p. 126).

However, circling back to the topic of this chapter, the interest in making a brief reference to the evolution of algebraic language is in pointing out that, despite the fact that the symbols and expressions of this language have their origins in contexts that are loaded with meaning, such as geometric and those coming from word problems, algebraic symbolism, as we know it today, is far from evoking referents to those contexts among current users. In fact, the power of this language is understood to reside in its generality, in the possibility of using it to model situations and problems from a variety of contexts, and that through transformations of the expressions or 'model' equations problems can be solved and knowledge of the modeled situations can be deepened. This possibility is precisely due to the fact that the symbols and expressions formed with them are not, from the outset, tied to particular or concrete referents. It is an autonomous, self-contained language, in which the utterance of the theorems and their proofs (Klein, 1968) and the statement of the problems and their solutions (Rojano, 1996) can be entirely expressed.

Along with the recognition of the unequivocal power of the abstract nature and generality of algebraic symbolism, one must acknowledge that teaching and learning said symbolism, its syntax, and its use represent a great challenge for mathematics education. The specialized literature on algebraic thinking reports a number of difficulties faced by students of secondary and tertiary education when they learn the rules of transformational algebra and, to a large extent, related studies attribute such difficulties to the lack of referents, to the lack of meanings with which the students can associate algebraic symbols, as well as to the absence of a sense they can develop with respect to the grammar of expressions and their transformations. Starting in the early 1980s, and still today, much research has been done on this issue, both to try to unravel the nature of the difficulties identified, on the one hand, and to propose and test different approaches to teaching algebra, on the other.

However, work carried out in the field of mathematics education on the intrinsic properties of the algebra sign system and its relationship with the users of that system has been much less abundant. The studies on structure sense by Hoch (2007) and Hoch and Dreyfus (2004) for symbolic algebra, and by Novotná and Hoch (2008) for abstract algebra, as well as the research done by Kirshner (1989) on the visual syntax of algebra set the stage for issues in that domain and leave open questions about the extent to which knowledge of this internal order of algebraic *texts* can lead to a better understanding of the difficulties faced by learners and to further opportunities for proposing new teaching approaches that take the inextricable subject-structure relationship into account.

Internal Order and Awareness of the Structure

An analysis of the structural properties of algebraic expressions is found in 'The visual syntax of algebra' by Kirshner (1989). In this article, the author considers a structured system of visual features in parallel with

6   *Teresa Rojano*

Table 1.1 Syntactic Convention Using Hierarchy of Operation Levels

Level 1	Addition	Subtraction
Level 2	Multiplication	Division
Level 3	Exponentiation	Finding Roots

(Level 3 is said to be higher than Level 2, which is higher than Level 1.)

Syntactic Convention

1. Higher-level operations have precedence over lower level operations.
2. In case of an equality of levels, the leftmost operation has precedence.

Taken from Kirshner, D. (1989). Visual syntax of algebra, p. 276. *Journal for Research in Mathematics Education*, May.

the parsing tree associated with an algebraic expression, which reveals the hierarchy of operations. Kirshner points out that despite the fact that the syntax rules, usually associated with the hierarchy of operations, are propositional in nature, a careful analysis of the algebraic notation shows that the levels of operation correspond to distinctive visual features (see Tables 1.1 and 1.2, taken from Kirshner, 1989, p. 276). That is, while in terms of propositional rules, higher-level operations have precedence over lower-level ones (i.e. exponents and radicals precede multiplication and division and these, in turn, precede addition and subtraction), in visual terms, this same criterion means that diagonal juxtaposition has higher precedence than horizontal or vertical juxtaposition. Additionally, Kirshner notes that, in parsing an expression, in addition to explicit symbols, such as parentheses, brackets, and braces, which force parsing according to their location in the expression, there are implicit parsing features, such as grouping by raising to a power (for example, at x^{3y}). In the latter case, in which there are no explicit markers, parsing determines the conventional hierarchy of operations.

Furthermore, the author argues that there is a strong correlation between the structured visual system of algebraic notation and the semantic categories (types of meanings) that underlie the propositional syntax (the rule-based syntax), which (he further argues) provides for the possibility that the syntactic rules be inferred in terms of visual

Table 1.2 Alternative Characterization of Operation Level

Level	Visual Characteristics	Examples
1	Wide spacing	$a + b, a - b$
2	Horizontal or vertical juxtapositions	$ab, \dfrac{a}{b}$
3	Diagonal juxtaposition	$a^b, \sqrt[a]{b}$

Taken from Kirshner, D. (1989). Visual syntax of algebra, p. 276. *Journal for Research in Mathematics Education*, May.

features directly, without intermediation by declarative representations. Kirshner discusses this and other hypotheses that are derived therefrom on the basis of the results from a study involving 211 students of 9th and 11th grades (reported in the same 1989 article), where the students are asked to evaluate expressions, such as $1+ 3x^2$ for $x = 2$ in standard notation (for example, $a + b$ or ab), unspaced nonce notation (for example, aAb or aMb), and spaced nonce notation (for example, $a A b$ or $a M b$), to analyze their ability to transfer the skills from numerical evaluation in standard notation to unspaced nonce and spaced nonce notations. Kirshner observed that most of the participants were able to correctly evaluate the expression in standard form, but found it much more difficult to transfer that ability to unspaced nonce notation than to spaced nonce notation. According to the author, this suggests that, for some students, the surface features of ordinary notation provide the necessary keys to make successful syntactic decisions (Kirshner, 1989, p. 282).

Along a different line, albeit with similar concerns, Hoch and Dreyfus (2004) carried out a study in which, by analyzing the results from applying a questionnaire to a group of high school students, they show that the presence of brackets and the position of the unknown variable or quantity in an equation affect the use of the ASS (the items used are shown in Figure 1.1). The definition taken as the starting point by the authors, which is applied to the high school level of education, describes

A. $$1 - \frac{1}{n+2} - \left(1 - \frac{1}{n-2}\right) = \frac{1}{110}$$

B. $$\left(1 - \frac{1}{n-1}\right) - \left(1 - \frac{1}{n-1}\right) = \frac{1}{132}$$

C. $$1 - \frac{1}{n+3} - 1 + \frac{1}{n+3} = \frac{1}{72}$$

X. $$\frac{1}{4} - \frac{x}{x-1} - x = 5 + \left(\frac{1}{4} - \frac{x}{x-1}\right)$$

Y. $$\left(\frac{1}{4} - \frac{x}{x-1}\right) - x = 6 + \left(\frac{1}{4} - \frac{x}{x-1}\right)$$

Z. $$\frac{1}{4} - \frac{x}{x-1} - x = 7 + \frac{1}{4} - \frac{x}{x-1}$$

Figure 1.1 Items used in the study on algebra structure sense with 11th grade pupils. Taken from Hoch and Dreyfus (2004, p. 51). Students were asked to solve the equations.

ASS as a collection of abilities, which include the ability to: see an algebraic expression or sentence as an entity, recognize an algebraic expression or sentence as a previously met structure, divide an entity into sub-structures, recognize mutual connections between structures, recognize which manipulation it is possible to perform, and recognize which manipulations it is useful to perform (Hoch & Dreyfus, 2004, p. 351). This definition is more general and less operational than the definition developed by Hoch in 2007 (stated at the beginning of the introduction to this chapter). Among the results obtained, it is reported that only 6.3% of the participants used ASS in the items without brackets, that is, for a majority, these markers were a factor that favoured the identification and cancellation of like terms. However, in item Z, which has no brackets, a non-explicit marker (the fraction line in $\frac{x}{x-1}$) seems to have had the same effect (Hoch & Dreyfus, 2004, pp. 3–53). The conclusions reached by the authors are that: (a) most of the participants did not use ASS in solving the equations; (b) those who did use it put into play the ability to see an algebraic expression or sentence as an entity, without applying algebraic transformations automatically, and the ability to recognize connections between structures; and (c) the latter led them to choose appropriate transformations to solve the equations (Hoch & Dreyfus, 2004, pp. 3–56).

With an approach that differs from that of Kirshner's, in this research Hoch & Dreyfus address the problem of unraveling the processes that take place during the interaction of subjects with the internal structure of objects in symbolic algebra, and they manage to identify factors that influence the awareness of said structure, such as the presence of explicit markers such as brackets (items B and Y in Figure 1.1) and the location of the unknown quantity in an equation to be solved (items X, Y, and Z in Figure 1.1). Kirshner, for his part, reports in his study that implicit markers, such as exponentiation (diagonal juxtaposition), can also be decisive in the perception of the structure of an expression. In addition, this researcher is of the opinion that other factors, such as processes of a cognitive nature and the way in which manipulative algebra is taught, can influence the way in which subjects perceive the structure and choose the transformations they perform. Although in his study the results are not conclusive in this regard, they are in the sense that this influence, in any event, does not occur homogeneously among all individuals.

The studies described above provide fundamental elements to raise questions in the field of didactics about the possibility of creating learning situations that promote development of ASS. That is, on the one hand, there are definitions of ASS, both of a general and abstract nature (Hoch & Dreyfus, 2004) and of an operational nature (Hoch, 2007), whose usefulness and scope have been tested in particular empirical studies. On the other hand, the works of Kirshner (1989) and Kirshner and Awtry (2004) identify a set of factors that influence the perception of the structure of algebraic objects, factors originating both from the structure of said objects (parsing of expressions), and the presence of elements of visual salience therein (elements of a structured visual system), as originating from individual differences of a cognitive nature and from types of initiation for the knowledge of algebra syntax. In the next section, we discuss the possibility of taking advantage of these results and results from other research in teaching and learning the syntax of algebra.

Algebra Structure Sense Development

In the case of Sonya L. (from Krutetskii's research), it is worth considering whether, through teaching, average students can develop this type of ability, which in Sonya is manifested as an awareness of the structure of a complex expression and the immediate application of a simple rule (the distributivity of the product with respect to the sum). To address this question, it bears remembering that the results of specialized ASS studies on visual aspects of algebraic syntax, such as those described above, point to a series of factors that can influence the development of ASS, which include the type of initiation for learning the syntax. In this regard, Kirshner distinguishes two types of approaches: syntactic knowledge initially acquired in propositional form (formal approach based on learning rules in declarative form) and syntactic knowledge initially acquired in a visual modality (training approach based on visual features that reveal structural aspects); and he formulates hypotheses about possible transitions from one type of modality to another (Kirshner, 1989, p. 275). In order to find possible answers to the question at hand in this way, it would be necessary to carry out more studies on the effects of these (and other) types of initiation on a subsequent development of ASS and relate them to individual differences, such as age, gender, or aspects of a cognitive nature, among others, which can lead to an interesting research agenda on teaching and learning manipulative algebra.

On the other hand, research in the field of early algebra opens up different avenues. Recent studies show that young learners can perform tasks that involve the knowledge and use of the structural properties of numbers and their operations, which is considered by several authors as fundamental background for developing algebraic thinking. While a good number of these research projects focus on studying the role of structural numerical thinking in the tasks of patterns and generalization, others focus on analyzing the potential that elementary school pupils have for working with additive structures of numbers and decomposition and re-composition of numbers.

Among the large number of studies on the first of these lines, the work of L. Radford (2011) on non-symbolic algebraic thinking stands out. The latter author argues that, although there is a growing body of experimental evidence to support the idea that an early introduction to algebra can give students access to advanced algebraic concepts in later grades, the idea lacks precision due to the lack of clarity on the distinction between what is arithmetic and what is algebraic (Radford, 2011, p. 304).

The argument put forth by this author has theoretical and empirical bases. On the one hand, in the empirical arena, it refers to the analysis of the data obtained in a study undertaken with 2nd grade children, in which, when carrying out an activity that involves extending a figurative sequence, the students demonstrate a coordination of spatial structural aspects, relative to the spatial arrangement of the elements of the figures in the sequence, with structural aspects of the numbers associated with those figures. The set of elements that make up the corpus of analysis of the study (writing, drawings, language, gesture), as well as the detailed analysis of it, allow Radford to identify manifestations of what he calls *non symbolic algebraic thinking*, when students try to answer questions about figures located in a remote location of the sequence, beyond the visual field, (removed from the possibility of proceeding by counting), and they show a tendency to think on the basis of unknown quantities as if they were known, without resorting to their representation with letters. At this point, the author resorts on a theoretical level to the notions of *indeterminacy* and *analyticity*, referring to the former as the condition of 'unknown' of the indeterminate quantities in the sequence, and to the latter, as the way to treat the unknown as if it were known.

Analyticity is then, according to Radford, a non-symbolic algebraic thinking trait, which, according to his results, can be observed in very young students when performing pattern generalization tasks. This conception of early algebraic thinking contrasts with others, where 'the algebraic' is associated with hallmarks of the presence of other types of algebraic concepts (such as a functional relationship) observed in student performances when solving the same type of task. The aforementioned empirical study and Radford's conception of what is 'algebraic' are particularly relevant to the subject of ASS, since both consider the coordination of structural aspects of the figurative and the numerical, as well as the role of analyticity in generalization processes. That is to say, from this perspective, the analyticity concerning unknown quantities in generalization tasks with figurative sequences can be interpreted as the algebraic element that links two coexisting structures, that of the figures in the sequence and that of the corresponding sequence of numbers. Despite the hallmarks found in his study, Radford points out that it is still too early to forecast the impact that early experience with pattern activities may have on the longitudinal development of the algebraic thinking of students in subsequent grades (Radford, 2011, p. 319).

On the second line, in which the potential of young students to work with additive structures of numbers and with the decomposition and re-composition of numbers is analyzed, we have the work of C. Kieran, which encompasses other numerical structural relationships (beyond additive relationships). Kieran addresses the issue of the notion of structure and structural activity in the field of the numerical from a study with 12-year-old pupils, who produced different structural decompositions of numbers in activities based on the problem 'Five Steps to Zero'[2] (Kieran, 2018). The decompositions that these students generated correspond to order structures, additive and multiplicative structures, as well as to a structure in which multiplication and division are combined using the division algorithm (Kieran, 2018, p. 22). The results obtained by Kieran et al. show that the design of the activities used and adapted to the age of the participants led to an evolution from trial-and-error strategies towards techniques linked to structuring strategies that involve the relationship of mutually inverse operations between multiplication and division.

One of the conclusions reached by this author is that the ways in which students used and expressed alternative structures when solving the problem 'Five Steps to Zero' show how students at this age (12 years) manage to grasp structural properties of multiplication and division in the context of a numerical activity (Kieran, 2018, p. 23) and that the experience with structuring activities like this one triggers forms of thinking that will be of great value for the structural understanding requirements of high school algebra. In this regard, it should be noted that the study described comprises two structural aspects: one, related to the internal structure of numbers, which allows decompositions and re-compositions of specific numbers; and another, concerning the structure of the number system (numbers and their operations), with strategies that relate to an understanding of the inverse operation relationship between multiplication and division being especially interesting. The above is clearly connected to the structural properties of algebraic operations, which play a fundamental role in learning algebra both at the conceptual level and in the development of syntactic skills.

The two studies referred to in this section provide significant data on the ability of young learners to grasp and use structural elements, whether numerical (in Kieran's study) or algebraic (the analyticity in the Radford study) in nature. Both contributions support the idea that early experience with these structural aspects constitutes very valuable background for development of ASS in secondary education. However, in accordance with one of the conclusions reached by Radford in his research, it should be noted that in specialized literature on early algebra in general there is a notable lack of strict longitudinal studies (following the same generation of students over the course of several years) that can account for such a development.

Technology, a Gateway to the Development of ASS

In his 1989 article (oft cited in the two previous sections), D. Kirshner points out that a curriculum that de-emphasizes the teaching of a propositional algebraic syntax implicitly delegates the responsibility to create conditions that will allow students to develop manipulative skills to the power of the visual traits of algebraic notation. In terms of ASS, this statement could be expanded to say that a teaching approach to algebraic syntax, without emphasizing its propositional component and without proper exercise, would mean that ASS development would depend on the effectiveness of the visual traits of algebraic expressions. In this respect, the results obtained empirically by Hoch and Dreyfus confirm that, indeed, there are markers that are visually effective for gaining awareness of the structure of an expression. Nevertheless, it would seem very risky to entrust development of manipulative skills to the visual structure of the algebra sign system *per se,* and then the question arises as to what alternatives can be designed to promote such development.

Beyond the interest that the previous question could have in the field of research, given its natural relationship with instruction, the question inevitably refers us to the old question raised in the context *of the so-called mathematics war* of the 1980s concerning the importance of teaching algorithmic skills and symbolic manipulation. The war of mathematics was a movement that consisted of contrasting the teaching of traditional contents with the guidelines of the reform initiated in the United States, in which priority is afforded to teaching concepts and solving problems, over algorithmic skills and symbolic manipulation (Van de Walle, 2007).[3] With interesting variations, that position that de-emphasizes such skills and that is expressed in the guidelines of the National Council of Teachers of Mathematics (NCTM) standards document (NCTM, 2000) has been adopted by a good number of educators and educational systems. Nonetheless for several years research has revealed that, despite the conceptual emphasis of such an approach, the understanding of mathematics concepts has not occurred as expected.

This educational trend that subordinates the algorithmic and symbolic fluidity has begun to no longer be the dominant approach, at least in some current curricular proposals, in which the value of basic technical aspects of mathematics is recovered (Van de Walle, 2007). While in that same direction, albeit in the context of research, it has been shown that students achieve conceptual progress when they engage in transformational algebra activities (see, for example, Solares and Kieran, 2013). Those results do away with the false dichotomy between learning concepts and learning symbolic skills (Kieran, 2013) that underlies that movement of the mathematics war.

Works that provide evidence that transformative algebra activities promote the comprehension of fundamental concepts (such as algebraic equivalence and substitution) include the use of computer algebra systems (CAS). Such is the case of studies carried out by Kieran and Saldanha (2008). In other words, incorporation of technology has been key to showing a possible confluence of the learning of conceptual aspects and of manipulative competencies.

When Kirshner's article in reference was published, the presence of technology usage in mathematics curriculum and teaching was very weak. Use of scientific calculators was probably what had most begun to be widespread in secondary and high school. However, as of the 1990s and to date in the field of research, numerical calculators have been used very frequently in various studies, some of which include pattern and generalization tasks as an aid in numerical calculations. An example of this is the Radford study described in the previous section. Moreover, and as already mentioned, calculators that include a CAS have played a key role in certain types of research, where the interest is studying conceptualization processes in algebra.

Other tools, other than calculators, have been tested creating learning environments for different aspects of algebra. One such tool is Spreadsheets, which have proved in studies conducted by Rojano and Sutherland to be an enabling environment for drawing students closer to algebraic ideas, such as function (Sutherland & Rojano, 1993) and quasi-algebraic methods of solving problems involving several variables (Yerushalmy & Chazan, 2008). In more recent studies, use of programmes such as GeoGebra has revealed the advantages of having multiple representations of concepts, dynamically interconnected, which in turn makes it possible to interconnect algebra, geometry, and the mathematics of variation.

Despite the incorporation of various tools for the teaching and learning of algebra, in research and curriculum teaching technology has been ill considered as a means of supporting fluency in symbolic manipulation. Chapter 4 discusses how more modern tools can contribute to the achievement of this type of competence and refer to the affordances provided by gamification and adaptive and artificial intelligence systems. In particular, by way of example, the *Expression Machine* web environment (*MEx*, for its acronym in Spanish) that was designed to promote ASS development is described.

One of *MEx*'s components is the design of a universe of exercises or tasks organized by levels of complexity, where complexity is defined, on the one hand, in terms of the parsing tree of algebraic expressions and, on the other, in terms of elements of visual salience. Another component is an adaptive system that allows each user, depending on her/his performance, to travel along paths of progress towards increasing levels of structure sense. Some gamification elements (in the form of challenges) are incorporated as part of the feedback system. One of the traits of the environment is that for many of the activities users must perform the transformations using paper and pencil algebra. That is to say, the tool is not a symbolic manipulation system; rather, it is a system that proposes challenges and means of verifying the result of the algebraic work performed 'manually'. In other words, it is a modern version of exercising in the area of symbolic manipulation, geared to development of structure sense.

In Chapter 5 of this book, Solares and Rojano discuss outcomes from a study with *MEx* conducted with a group of users with different mathematical backgrounds in the context of the works on structure sense discussed here. Chapter 5 also presents arguments that support the hypothesis on the feasibility of heterogeneous groups of users evolving towards higher levels of ASS, by way of interacting with an adaptive system organized by levels of complexity of algebraic objects and using elements of visual salience of those objects. This also strengthens the idea of a renewed version of algebraic exercise, which is far away from the traditionalist notions of skillfulness and drill and practice that were so in vogue in the 1950s and early 1960s in the field of learning theories.

The Importance of ASS in Advanced Mathematical Thinking

Symbolic manipulation skills have a clear and direct application in solving differential and integral calculus tasks, the techniques and methods of which involve transformations of analytical expressions of the functions. If one assumes that development of such manipulative skills depends largely on an adequate ASS development, then the latter can also be considered a direct and important antecedent for successful performance in tertiary education calculus courses. One can also refer to other areas of mathematics for which a good mastery of symbolic manipulation and, therefore, a good level of ASS development are necessary basic capabilities. However, the question is whether a good development of ASS influences the understanding of concepts or performance in other areas that are not so clearly related. The study by Novotná and Hoch (2008) answers that question in the specific case of abstract algebra. Considering that each of the structures of abstract algebra (group, ring, field) is associated with a closed set under one or more operations that satisfy some axioms, the latter authors characterize the structure sense of abstract algebra, on the one hand, in reference to the elements of the sets and the notion of binary operation and, on the other, referring to the properties of the binary operation.

In terms of the set and notion of binary operation, Novotna and Hoch say that students have structure sense if: (a) they recognize a binary operation in familiar structures; (b) recognize a binary operation in non-familiar structures; (c) see elements of the set as objects to be manipulated, and understand the closure property. While in terms of the properties of the binary operation, they are of the opinion that students have structure sense if: (a) they understand the identity element in terms of its definition; (b) see the relationship between identity and inverse elements; (c) use one property as a supporting tool for easier treatment of another (e.g. commutativity for identity element, commutativity for inverse element, commutativity for associativity); and (d) keep the quality and order of quantifiers. Based on these definitions, the authors conducted a study in which they sought to investigate how structure sense of algebraic expressions or equations (corresponding to high school algebra (HS)) is related to structure sense in abstract algebra (corresponding to university algebra (UA)). To accomplish this, they applied a questionnaire with four tasks to a group of pre-service mathematics teachers. Two of the tasks belonged to HS algebra: (1) Solve in R: $(x+3)^2 \leq 6x+18$; (2) Simplify in R: $\frac{(x-2)^4-(x-2)^4}{14x^2-2x}$. While the other two belonged to UA: (3) Let $x \Delta y = \frac{9x^2-16y^2}{6x-8y}$, decide whether Δ is a binary operation on the set of real numbers and if so, determine its properties; (4) Consider the following structure $(Z, \cdot)$, where Z is the set of integers, and $x \cdot y = x+y-4$. If $\cdot$ is a binary operation on Z, determine its properties. If the neutral element exists, find the inverse elements of all integers for which they exist (Novotná & Hoch, 2008, p. 100).

Although the structure sense of HS algebra and the structure sense of UA seem to be two unrelated worlds (as according to Novotna and Hoch themselves the former is based on symbolic thinking and the latter pertains to the world of formal thinking), the results obtained suggest that the HS symbolic world structure could be a prerequisite for the structure sense of the formal world of UA (Novotna & Hoch, 2008, p. 102), despite the fact that their study does not provide elements that make it possible for us to speak of the nature of a possible connection or of a transition from one structure sense to another. Chapter 7 provides a broader discussion of the possible ASS connections with other mathematics domains, not necessarily of an algebraic nature, in order to demonstrate the relevance of ASS development in the thinking of advanced mathematics.

Connection to Upcoming Chapters

In the previous sections, different perspectives and research were presented and analyzed, perspectives that contribute theoretical and empirical elements to the study of the notion of ASS, and topics that are developed more extensively in the upcoming content of the book were discussed. Thus in the interest of expanding upon the brief reference made to the evolution of algebraic symbolism and of contributing to reflection on the nature of structural aspects of different algebra sign systems, in Chapter 2 Luis Puig analyzes how the history of the idea of structure in modern algebra, particularly how the structural properties of sets of objects in this domain are the result of abstraction processes that took place throughout a history of seeking, using, and studying structural properties in arithmetic and in solving symbolic algebra problems.

As pertains to the approach put forward on the possibility of creating conditions in elementary school for the development of ASS that would allow students access to algebraic knowledge in secondary school, in Chapter 3 Kieran and Martinez present and discuss results of a study, in which 6th grade students show how the strategies they use in numerical equivalence and equality activities evolve from computational strategies to structural strategies.

In terms of the role that technology may have in reformulation of the algebraic exercise, in Chapter 4 Muñoz and Xolocotzin discuss how current developments in interaction design such as adaptivity and gamification can be used to support algebraic processes, such as the development of structure sense. Additionally, these authors introduce the *MEx* platform, designed to promote ASS integrating theoretically guided instructional and representative principles with components of adaptivity and gamification. In Chapter 5, Solares and Rojano use the results obtained in the *MEx* experimental work with users with varied mathematical background, as a departing point to study from a semiotic perspective the production of meanings during the processes of appropriating the structure sense of algebraic *texts*.

A previous section also refers to a number of universal technological tools, such as Spreadsheets, CAS, and Geogebra, which have been widely used for teaching and learning algebra. In Chapter 6, Geraniou and Mavrikis present a study with children aged 11 to 14, using microworld *eXpresser*, which, unlike the programmes mentioned, is an ad-hoc development that consists of a toolkit for building tilling patterns and includes a collaborative and reflection phase. The main purpose of this study is to close the gap between identifying and expressing patterns, encouraging students to identify the structure of the pattern by building it. In that chapter, the authors discuss the results that deal with the students' justification strategies concerning the correct nature of their constructions and, where appropriate, regarding the equivalence of the rules they formulated using the tool. Since *eXpresser* is an environment for working with diagrammatic, numerical, and symbolic representations of patterns, the content of this chapter contributes to deepening Radford's approaches to the coordination of the unlike-natured structures referred to in Section 4.

Chapter 7 by Rojano and Palmas is dedicated to making explicit mathematical abilities that correspond to different educational levels, abilities that are related to the development of ASS, and that contribute to the mathematical maturity of individuals. For that purpose, we reflect on the sequencing of topics within curricular organizations that suggest the prerequisite condition of ASS development for access to the mathematics content of post-secondary levels. Additionally, the role of ASS in development of advanced mathematics skills or in the learning of higher education mathematics content is analyzed. In both cases, the authors refer to concepts, ideas, and positions discussed in the present chapter.

Finally, Chapter 8 provides an account of how Chapters 1–7 contribute to addressing the need to broaden the meaning of the term *algebra structure sense*. It is for that purpose that the authors resort to elements provided by different perspectives, related both to features of the internal structure of algebra objects and to cognitive and semiotic processes that take place in the perception of the objects or during their manipulation. Likewise, the author re-broaches the topic of the role of modern technological developments in reformulation of the so-called algebraic exercise, in terms of the development, by levels of complexity and performance, of the structure sense.

Notes

1 As for the fundamental rules for setting up the equations, the art of Zetetics is embraced in the art of logic. It does however have its own method or mode of proceeding as it is not limited to reasoning with numbers (characteristic of the old analyses), rather in order to compare magnitudes it works with a new symbolic logic that is more efficient and powerful than numerical logistics (Witmer, 1983).

   Setting up an equation involves working with elements of symbolic logistics or specious logistics, namely elements called *species* that substitute numbers. Since *species* have a more general nature than *arithmos* and *eidos* (names for known quantities), those elements require a redefinition of operations between them. It is for this reason that Viéte explicitly states the basic laws of Zetetics, under which *species* are assigned dimensions and the restriction of equidimensionality (homogeneity) is imposed for setting up any equation so that two expressions can be compared (i.e., so that they can appear in the same equation) (Klein, 1968).

2 Five steps to zero problem consists of asking pupils to take any whole number from 1 to 1000 and try to get it down to zero in five steps or fewer, using only the whole numbers 1 to 9 and the four basic operations. The same number may be used more than once in the operations (Kieran, 2018, p. 12).

3 *Mathematics war* is the term given to the debate in which the teaching of traditional content was confronted with the guidelines of the reform begun in the 1980s in the USA, in which priority is given to teaching concepts and problem-solving above algorithmic skills and symbolic manipulation (Van de Walle, 2007).

References

Cajori, F. (1928–1929). *A history of mathematical notations, 2 vols*. Chicago: Open Court.

Eves, H. (1983). *Great moments in mathematics, before 1650*. USA: The Mathematical Association of America.

Esty, W. W. (1992). Language concepts of mathematics. *Focus on Learning Problems in Mathematics*, 14(4), 31–53.

Heath, T. L. (1910). *Diophantus of Alexandria*, rev. ed. New York: Cambridge University Press.

Hoch, M. (2007). Structure sense in high school algebra. *Unpublished doctoral dissertation.* Tel Aviv University, Israel.

Hoch, M., & Dreyfus, T. (2004). Structure sense in high school algebra: The effect of brackets. *Proceedings of the 28th PME Conference, 3,* 49–56.

Hughes, B. (Ed.) (1981). Jordanus de Nemore. *The numeris datis.* Berkeley, CA: University of California Press.

Kieran, C. (2013). The false dichotomy in mathematics education between conceptual understanding and procedural skills: An example from algebra. In K. Leatham (Ed.), *Vital directions in mathematics education research* (pp. 153–171). New York: Springer.

Kieran, C. (2018). Seeking, using, and expressing structure in numbers and numerical operations: A fundamental path to developing early algebraic thinking. In C. Kieran (Ed.), *Teaching and learning algebraic thinking with 5- to 12-year-olds: The global evolution of an emerging field of research and practice* (pp. 79–105). New York: Springer.

Kieran, C., & Saldanha, L. (2008). Designing tasks for the co-development of conceptual and technical knowledge in CAS activity: An example from factoring. In K. Heid, & G. W. Blume (Eds.), *Research on technology and the teaching and learning of mathematics: Syntheses, cases, and perspectives* (pp. 393–414). Greenwich, CT: Information Age Publishing.

Kirshner, D. (1989). The visual syntax of algebra. *Journal for Research in Mathematics Education, 20*(3), 274–287.

Kirshner, D., & Awtry, T. (2004). Visual salience of algebraic transformations. *Journal for Research in Mathematics Education, 35*(4), 224–257.

Klein, J. (1968). *Greek mathematical thought and the origins of algebra.* Cambridge, MA: MIT Press (*Reprinted* in 1992, New York: Dover).

Krutetskii, V. A. (1976). *The psychology of mathematical abilities in schoolchildren.* Edited by J. Kilpatrick and Translated from the Russian by Joan Teller. Chicago, London: The University of Chicago Press.

Mazzinghi, M. A. (1967). *Trattato di Fioretti [Fioretti's treatise].* In G. Arrighi (Ed.), Pisa, Italy: Domus Galileana.

NCTM (2000). Principles and standards for school mathematics, grades 9–12. Reston, VA: National Council of Teachers of Mathematics.

Novotná, J., & Hoch, M. (2008). How structure sense for algebraic expressions or equations is related to structure sense for abstract algebra. *Mathematics Education Research Journal, 20*(2), 93–104.

Radford, L. (2011). Grade 2 students' non-symbolic algebraic thinking. In J. Cai, & E. Knuth (Eds.), *Early algebraization* (pp. 303–322)). Berlin: Springer-Verlag.

Rojano, T. (1985). De la aritmética al álgebra. Un estudio clínico con niños de 12 a 13 años de edad (From arithmetic to algebra. A clinical study with children of 12 to 13 years of age). Unpublished doctoral dissertation, Centro de Investigación y de Estudios Avanzados (Cinvestav), Mexico.

Rojano, T. (1996). The role of problems and problem solving in the development of algebra. In N. Bednarz, C. Kieran, & L. Lee (Eds.), *Approaches to algebra. Perspectives for research and teaching* (pp. 55–62). Dordrecht, Boston, London: Kluwer Academic Publishers.

Sfard, A., & Linchevski, L. (1994). The gains and the pitfalls of reification – The case of algebra. *Educational Studies in Mathematics, 26,* 191–228.

Sigler, L. E. (2002). *Fibonacci's liber abaci. A translation into modern English of Leonardo Pisano's book of calculation.* New York, Berlin, Heidelberg: Springer.

Solares, A., & Kieran, C. (2013). Articulating syntactic and numeric perspectives on equivalence: The case of rational expressions. *Educational Studies in Mathematics, 84*(1), 115–148.

Sutherland, R., & Rojano, T. (1993). A spreadsheet approach to solving algebra problems. *The Journal of Mathematical Behavior, 12*, 351–383.

Van de Walle, J. A. (2007). *Reform vs the basics: Understanding the conflict and dealing with it.* Virginia: Commonwealth University.

Witmer, T. R. (tr.) (1983). *The analytic art by François Viéte (1540–1603).* Kent, Ohio: The Kent State University Press.

Yerushalmy, M., & Chazan, D. (2008). Technology and curriculum design: The ordering of discontinuities in school algebra. In L. English et al. (Eds.), *Handbook of international research in mathematics education* (2nd ed., pp. 806–837). New York: Routledge.

2 Ideas of Structure in the History of Algebra and Its Teaching

Luis Puig

Problems, Equations, Structures

Structures and Ideas of Structure

Broadly speaking, algebra has focused first on the study of problems and their resolution, then on the study of equations and their resolution, and finally on the study of structures. To be more precise, it first focused on quantitative problems and their algebraic resolution, then on polynomial equations and their algebraic resolution, and finally on algebraic structures – and what "algebraic" means in "algebraic resolution" and "algebraic structure" needs to be specified. However, to say that the object of study of algebra in the latter part of its history is algebraic structures does not mean that ideas of structure did not previously exist and were studied.

According to Leo Corry, this focus of the study on structures can be dated back to 1930 when Van der Waerden's *Modern Algebra* was published. In Van der Waerden's textbook, "The structural image of algebra [...] is based on the realisation that a certain family of notions (that is, groups, ideals, rings, fields, etc.) are, in fact, individual instances of one and the same underlying idea, namely, the general idea of an algebraic structure, and that the aim of research in algebra is the full elucidation of those notions" (Corry, 2007, p. 222). This change of approach involves the reification of the means of organization of the phenomena studied when dealing with the objects of study on which algebra focused previously. Groups, for instance, were developed in order to study the resolution and the solvability by radicals of polynomial equations. Therefore, groups were developed and studied previously as a tool and not as an object, and as such "Groups [...] had appeared in mainstream textbooks on algebra as early as 1866, in the third edition of Serret's *Cours d'algèbre supérieure*" (Corry, 2007, pp. 222–223). Even in the introduction of the 1877's fourth edition of his *Cours*, Serret clearly stated that "Strictly speaking, Algebra is the Analysis of Equations; the various partial theories that it includes all relate, more or less, to this main object"

DOI: 10.4324/9781003197867-2

(Serret, 1877, p. 1). Serret, therefore, did not see structures as the object of study of algebra, but rather equations, despite the fact that he presented the structure of group in his textbook. However, the idea that algebra is focused on the study of equations, which Serret still asserted in 1877, is not the first on the object of study of algebra, nor could it be.

Algebraic Problem Solving

For a long time, algebra was presented as having the aim of solving problems. At the beginning of his 9th century foundational algebra book on *Calculation by al-jabr and al-muqābala*, al-Khwārizmī says, "I wanted it to enclose what is subtle in the calculation and what is most noble in it, what people necessarily need in their inheritances, their bequests, their distributions, their arbitrations, their businesses" (Rashed, 2007, p. 95 of the Arabic, p. 94 of the French translation). That is, algebra is what people need to solve problems in which one has to calculate with quantities (of money, in all the examples given by al-Khwārizmī[1]). Seven centuries later, at the end of the 16th century, Vieta explicitly says in the final paragraph of his programmatic book *In artem Analyticem Isagoge* (*Introduction to Analytical Art*) that the aim of algebra, which he calls "the Analytical Art", is "the problem of problems, which is: To leave no problem unsolved [Nullum non problema solvere]" (Vieta, 1591, fo. 9r).

However, problem solving is a general aim of mathematics, and algebra deals with a class of problems, quantitative ones, in which some unknown quantities have to be found from other quantities that are known by using a network of relations among them. Furthermore, what is specific to algebra is not the particular class of problems that it aims to solve, but the way it does so. As we said in Puig and Rojano (2004), reformulating Mahoney (1971), one of the characteristics of the algebraic is the "use of a system of signs to solve problems, which allows us to express the content of the statement of the problem relevant to its solution (its 'structure'), separated from what is not relevant to its solution" (Puig & Rojano, 2004, p. 198). Problems are solved through analysis and translation of their statement to expressions in that sign system, a translation that ends when a quantity has been expressed in two different ways, which can then be equated. That is, the translation ends with the writing of an equation, which expresses the content of the statement of the problem relevant to its solution (its "structure").

This algebraic way of solving problems, therefore, transfers the algebraic project of solving all problems to the problem of solving all equations. But for this, a sign system specific to algebra must have been elaborated in which to express that structure of the problems, the equations. Further, a calculation must have been developed in that sign system, which can be performed without resorting to the contextual meaning in the statement of the problem. The symbolic sign system taught nowadays in school algebra is such a sign system, and its development is the result of a long and nonlinear history of algebraic sign systems. In the last section of this chapter, we give an overview and a few snapshots of that history.

Al-Khwārizmī shows, through the organization of his foundational book, this transfer to solving all the equations from the algebraic project of solving all problems, and the key role that the elaboration of a specific sign system plays. Indeed, although in the prologue he states that the aim of algebra is to solve problems, the book does not begin by solving problems nor does it discuss how to solve a problem.

Instead, the book begins by referring to the forms that the numbers "needed in calculation" take when solving quantitative problems, because the calculation "of *al-jabr* and *al-muqābala²*" is not the arithmetic calculation performed with numbers, but rather a calculation that is carried out with the forms that the numbers take, which al-Khwārizmī calls "species of numbers". "I have found that the numbers needed in calculation by *al-jabr* and *al-muqābala* are of three species, namely: *jidr* (root), *māl* (treasure, wealth) and *cadad mufrad* (simple number) not related to root and treasure" (Hughes, 1986, p. 233; Rashed, 2007, pp. 96–97)³.

The "species of numbers" are the objects of algebraic calculation, because as they are "forms of numbers" rather than numbers, they allow us to represent and calculate with the unknown and not only with the known. The species of numbers is not an absolute property of a quantity, but a property relative to the calculations that are being performed with this quantity.

Al-Khwārizmī later introduces another technical term: the thing. This term is used to name an unknown quantity in a problem so as to be able to calculate with it, thereby establishing the species of numbers that it is. Usually the thing (name of an unknown quantity) turns out to be a root (species of numbers), which caused "thing" and "root" to sometimes be used interchangeably in the history of Arabic algebra, although the discussion about their difference can be found in Arabic texts from several centuries after al-Khwārizmī's book (Souissi, 2001).

Equations are equalities between aggregates (Oaks, 2012) of species of numbers, for instance "four treasures and nine dirhams less four roots equals to one root", and al-Khwārizmī, in order to be able to solve all equations, looks for all the possible combinations of the three species of numbers and finds that there are six: treasure equals roots, treasure equals numbers, roots equals numbers, treasure and roots equals numbers, treasure and numbers equals roots, and roots and numbers equals treasure.

The calculation of *al-jabr* and *al-muqābala* is intended to transform any equation into one of those six canonical forms. Now, all that is required is to find an algorithm for solving each of the canonical forms and the problem of solving all the equations composed of these three species of numbers will be solved.

Algebraic Equation Solving

Algebraic equation solving means a specific way of solving an equation, which a few centuries later Serret defined as follows:

> *The algebraic resolution of equations*, that is to say the determination of an expression composed with the coefficients of a given equation, and which, substituted for the unknown, identically satisfies this equation. (Serret, 1877, p. 1)

Al-Khwārizmī did not have an algebraic sign system in which it was possible to express such a general formula, but he had equivalent algorithmic procedures, shown in generic cases, to solve all six canonical forms. He was thus able to algebraically solve all equations made with the first three species of numbers.

Three centuries later ᶜUmar al-Khayyām attempted to solve all equations made with the first four species of numbers, adding the cube to the first three. He dedicated his *Treatise on algebra* exclusively to equation solving, without including any problem. He looked for all possible combinations of the four species of numbers and found that there were 25, but he could not find an algebraic solution of all 25 canonical forms. In the cases that eluded him, he showed how a solution could be constructed by means of intersecting conic sections. The algebraic solution of these canonical forms remained open until the 16th century.

We will not delve into the story of the difficulties and failures inherent in attaining algebraic solutions to all equations. What is of interest to us is highlighting that the difficulties and failures resulted in the equations themselves being taken as an object of study, and that the study of equations from the perspective of the conditions of their algebraic solubility (by Lagrange, Abel, and Galois) led to modern algebra[4].

Instead, we are going back in time to look at a beginning of algebraic problem solving.

Algebraic Features in Old Babylonian Problem Solving

The clay tablet TSM XIII, found at Susa, dating from the late Old Babylonian period (2000–1600 BCE), contains the statement of a word problem:

2 gur 2 pi 5 ban of oil I have bought.

From the buying of 1 shekel of silver 4 silà, each (shekel), of oil I have cut away.

2/3 mina {...} of silver as profit I have seen.

Corresponding to what have I bought and corresponding to what have I sold?

(Høyrup, 2002, pp. 206–207)

Høyrup's translation, which he calls "conformal", aims to "preserve the structure" (Høyrup, 2002, p. 41). Technical terms are translated in a manner that maintains the distinctions of their use in the historical text. Furthermore, the syntactic structure of the sentences is maintained even at the cost of violating the syntax of the language into which they are translated. The intent is for the reader to get the fundamental features of the conceptualization of the mathematical objects and processes involved in the original text and in the practices of that time. A paraphrase of Høyrup's translation in which we have brought the text closer to the English language, we have substituted the Babylonian metrological units, and we have written the numbers in the decimal number system is as follows:

A merchant buys 770 liters of oil. He sells them giving 4 liters less than what he has received in the purchase, for each gram of silver. He makes a profit of 40 grams of silver. Find out how many liters per gram of silver he bought the oil for and how many liters per gram of silver he sold it for.

Our paraphrase does not preserve the structure that Hoyrup wants to preserve with its conformal translation, but it does preserve the structure of quantities and relations that is relevant to the algebraic solution of the problem. Specifically, the quantities "amount bought and sell" (a, 770 liters), "purchase rate" (r_p, l/g), "rate of sale" (r_s, l/g), "amount of the purchase" (a_p, g), "amount of the sale" (a_s, g), "unitary profit" (p_u, 4 l/g), and "total profit" (p_t, 40 g), which are linked by the relations $a_p \times r_p = a$, $a_s \times r_s = a$, $p_u = r_p - r_s$, and $p_t = a_s - a_p$. Notice that as the price is expressed in quantity of product per unit of money, instead of quantity of money per unit of product, the relationship between the total amount of the purchase and that of the sale and the prices of the purchase and sale is inverse proportionality.

After the statement of the problem, the tablet contains its solution expressed as a sequence of operations, without any further explanation. The first operations, paraphrased from the text translated by Høyrup, are as follows (we have numbered the lines for easy reference):

1 You, 4 liters of oil posit and 40 grams the profit posit.
2 Inverse of 40 detach. 0,025 you see.

3   0,025 to 4 raise. 0,1 you see.
4   0,1 to 770, the oil, raise. 77 you see.
5   1/2 of 4 break. 2 you see.
6   2 square, 4 you see.
7   4 to 77 append. 81 you see.
8   What is equalside? 9 is equalside [...]

In lines [1] to [3], the unitary profit (4 l/g) is divided by the total profit (40 g). Division in Old Babylonian arithmetic is done by multiplying the inverse of the divisor (calculated in line [2]) by the dividend (line [3]). In line [4], the result of this division is multiplied by the quantity of oil bought and sold (770 l). None of these operations makes sense in the context of the story as told in the statement of the problem. Furthermore, from the rest of the operations it follows that the result of the multiplication carried out in line [4] turns out to be the product of the purchase rate by the rate of sale, which does not make sense either. Indeed, the rest of the operations follow the standard procedure in Old Babylonian mathematics to solve the problem of finding the length and width of a rectangle whose area and the difference between its sides are known, applied to a rectangle of area 77 (the quantity calculated in lines [1] to [4]) and difference between its sides 4 (the unitary profit). Once the results have been calculated, 11 for the length and 7 for the width, the tablet indicates that those are the values of the quantities that were to be found: "11 liters each gram you have bought. 7 liters each gram you have sold".

Lines [5] to [8] are the beginning of this standard procedure, known as the "Akadian method", in which the excess of the rectangle's length over its width is cut in half (line [5]), and one-half is moved and pasted on the other side of the rectangle, forming a gnomon. This gnomon is completed to a square (lines [6] and [7]), so that the area of the resulting square is known and, therefore, its side can be calculated (line [8]). From that side, the length and width of the initial rectangle are calculated by adding and subtracting half of the excess. All of these operations make sense as operations on a rectangle, but again they are meaningless in the commercial context of the problem.

We have shown in detail in Puig and Navarro (2010) how operations [1] to [4] make sense in a configuration of rectangles about the diagonal of a rectangle and its complements. In such a configuration, the sides of the rectangles about the diagonal and the rectangle are directly proportional and the sides of the complements are inversely proportional, since the complements are equal in area.

As Høyrup (2002) has extensively argued, geometrical configurations of rectangles, squares, and "broad lines" (rectangles with width 1) are a standard representation for problems, and their components are "functionally abstract". A segment may represent a number from a table of inverses, an area (as in BM 13901 #10), a volume (as in TMS XIX #2) or a commercial rate, as in TMS XIII, which we discuss here (Høyrup, 2002, p. 280). Problems that deal with other quantities are solved by first translating them to a geometrical configuration of this kind that is seen as having the same structure relevant to its solution. And then, by means of cut and paste and scaling operations (as in lines [1] to [4]) that make sense in the geometrical configuration, one looks for a configuration that one already knows how to solve. The known solution procedure of this last configuration, which makes sense in the geometrical configuration, gives the results in terms of the elements of the representation (length and width of a rectangle), which are translated back into the context of the problem statement (purchase rate and rate of sale).

Translating the statement of the problem into a kind of representation with the same structure; operating on the representation without having to refer back to the semantic context of the problem statement; transforming the representation of the statement of the problem into a representation one knows how to solve; translating the results back to the problem: these are all features of algebraic problem solving.

Building a Language for Algebraic Problem Solving: A Rough Overview and Some Snapshots

We have seen in the previous section that the kind of geometrical configurations used in Old Babylonian algebraic problem solving served as a sign system into which quantitative problems are translated in a manner that preserves the structure relevant to their solution, and that the problems were solved by performing operations that make sense in the semantic field of this sign system. These geometric configurations have two fundamental features of the algebraic language: (1) one can operate directly on them and (2) only what is pertinent to the solution of the problem (its structure) is translated into them. However, Høyrup (2002) points out that "Old Babylonian 'algebra' fails indubitably" in terms of one of the characteristics of algebra established by Mahoney (1971): "the search for relationships (usually combinatory operations) that characterize or define that structure or link it to other structures" (Høyrup, 2002, p. 281). No such search is found in extant Old Babylonian algebra tablets, and we dare say that the type of sign system used is not adequate or efficient for that purpose.

In any event, it turns out that the sign systems with which this search began to be carried out were developed from the previous conceptual elaboration of the "species of numbers" as basic objects of algebra. As far as we know, the first extant example we have of such a search is in al-Khwārizmī's book on algebra, in which the six canonical forms of equations are presented as all the combination possibilities in an equation of the three species of numbers that al-Khwārizmī considers, and in which the calculations by *al-jabr* and *al-muqābala* aim to transform any equation to one of these canonical forms, these six types of problem structures.

In his algebra book, al-Khwārizmī wrote species of numbers, algebraic expressions, and equations by means of words of natural language, Arabic words. Numbers are also written in words, despite the fact that he wrote another book explaining the decimal positional numeral system, in which the numerals referred to today as Arabic are introduced. The algebraic sign system taught currently in school algebra combines letters with numbers and some special signs, without using natural language words. This system has hardly undergone significant changes since Euler's 1771 textbook *Elements of Algebra*[5]. There is no global linear history of algebraic language that leads to this consolidated sign system. However, relationships can be established between this and other systems of signs developed throughout history. To do so, the characterization made by Nesselmann (1842) almost two centuries ago of three "stages in the development" of algebra can serve as a rough overview. We will also give some finer grain details in the snapshots that follow in this section.

Nesselmann's Types of Algebraic Language

Nesselmann used the way the "formal representation of algebraic equations and operations" was made as the criterion to divide the history of algebra into three stages. These stages are then based on the distinction of different types of algebraic sign systems.

In the first, algebraic expressions and equations and all operations on them are expressed by means of words of natural language: this stage is called "rhetoric". This is the case of an equation the likes of "four treasures and nine dirhams less four roots equals to one root", its transformation by *al-jabr* into "four treasures and nine dirhams equals to five roots", and its further reduction to one treasure in order to identify it with one of the canonical forms. Equations and operations are expressed in al-Khwārizmī's algebra completely in words, as in most of Arabic algebra. The same happens with medieval algebra in the Christian West, for example, in the final part of Leonardo de Pisa Fibonacci's *Liber Abbaci*, which is devoted to algebra (see Giusti, 2020, pp. 622–690).

The second stage is called "syncopated", and it differs from the rhetoric stage only in that "for certain frequently recurring concepts and operations it uses consistent abbreviations instead of complete words" (Nesselmann, 1847, p. 302). In Arabic algebra, this is the case of a sign system developed in the Maghreb and al-Andalus in which the names of each species of numbers were represented by its first letter, the word "less" by its two last letters and the word "equals" by its last letter, and to write a polynomial or an equation the abbreviations for the species of numbers were written above each number with numerals (see Abdeljaouad, 2002, and Oaks, 2012). This development in Arabic algebra was not known when Nesselmann wrote his book, in which he mentioned as syncopated algebra "Diophantus and the later Europeans up to the middle of the seventeen century" (Nesselmann, 1847, p. 302). Diophantus abbreviated also the names of his "species of numbers" (*monas, arithmos, dynamis,* and so on) by taking the two first letters of each, and combining them (putting the "o" above a capital "M" for *monas,* raising the "y" after a capital "D" for *dynamis*[6], for instance), and he also has an abbreviation for "less" and for "equals". As an example of the abbreviations used in Europe, the rhetoric equation "four treasures and nine dirhams less four roots equals to one" would have been written in the most popular textbook on arithmetic and algebra printed in Spain, Pérez de Moya's *Arithmetica practica y speculativa,* published in 1562, and reprinted nearly 30 times, as:

4 *ce p* 9 *n m* 4 *co ig.* à 1 *co*

Abbreviations are made by using the first or the first two letters of the words, but it is worth noting that "ce" abbreviates "censo", the Spanish translation of the Latin word "census" that Gerardo de Cremona used to translate the Arabic *māl,* "treasure"; and the same is done with "n" (for "número", number), "co" (for "cosa", thing, identified with root), and "ig. à" (for "igual à", equal to). But the words in Spanish for "plus" ("más") and "minus" ("menos") begin both by "m", and Pérez de Moya borrows the abbreviations "p" and "m" from Italian algebra, where Pacioli or Tartaglia had used them. To a Spanish speaking reader "p" is not an abbreviation of a known word, but rather it is a sign whose meaning is established by convention[7]. In fact, Pérez de Moya introduces them by simply stating "la p quiere decir mas, y la m menos (the p means plus, and the m minus)" (Pérez de Moya, 1562, p. 453).

The German version of the syncopation of algebraic language includes a set of signs to represent the species of numbers that were called "cossic signs" (from "Coss" the German translation of "thing", from the italian "cosa"). These cossic signs at first sight seem like conventional signs, symbols. However, according to Cajori, the cossic signs that were used in German manuscripts before they appeared in print are schematic abbreviations: "Paleographers incline to the view that [the sign for the thing[8]] is a modification of the Italian *co,* the o being highly disfigured" (Cajori, 1993, vol. I, p. 338). The previous equation would be written with cossic signs like this:

4𝖝 + 9𝕼 – 4𝖗 *gleich* 1𝖗

It is not difficult to recognize the first one as the letter "z", the first letter of "Zensus", the German translation of Latin "census". But it's true that "co" has been highly disfigured in the sign for "Coss", thing. Note also that German algebraists did not use abbreviations for plus and minus, but the signs + and –. Actually, "+ sign descended from one of the florescent forms for *et* in Latin manuscripts" (Cajori, 1993, vol. I, p. 231), that is, + is a stylized abbreviation of a word used in Latin for addition that is no longer seen as an abbreviation, but rather as a conventional sign.

Finally, the third stage is called "symbolic". Nesselmann's characterization of what symbolic means implies that expressions and algebraic equations are not written with words or with word abbreviations, but with other signs. Now, this is not what Nesselmann emphasizes as the fundamental characteristic of the symbolic, but rather that operations with algebraic expressions can be carried out without the need to resort to words.

> We can perform an algebraic calculation from start to finish in a wholly understandable way without using a single written word, and, at least in comparatively simple calculations, we really only place a conjunction here and there between formulae so as to point directly to the connection between a particular formula and those that precede and follow it, in order to spare the reader the need to search and reread.
>
> (Nesselmann, 1842, p. 302)

Vieta is known as the father of symbolic algebra. However, his sign system for *logistica speciosa*[9], the name he gave to his new algebra, is not symbolic in Nesselmann's sense. In his novel sign system, Vieta introduced letters to represent unknown and known quantities, so that in his algebraic expressions not only the species of numbers are represented, but also the quantities. This enabled him to express equations and transformation rules in a general way. However, he left in words (or sometimes abbreviated) the names of the species, the multiplication (Latin word *in*) and the equality (Latin word *aequetur*). The above equation would be written in Vieta's *logistica speciosa* as something like this:

A quad in B + C solido – A in D plano aequetur A in Z plano

where the consonants B, C, D, and Z generalize the known quantities 4, 9, 4, and 1, and the vowel A represents the unknown quantity.

To find an algebraic sign system symbolic in the Nesselmann sense, one has to wait until further developments in which the multiplication of quantities is expressed by the juxtaposition of the letters representing them (as was done by Harriot, after Vieta) and the names of the species are replaced by a number (as was done by Chuquet and was done anew by Bombelli, before Vieta). The combination of these developments results in a sign system that is efficient for calculating directly on the expression level following rules, that is, symbolic in Nesselmann's sense, such as that elaborated by Descartes or that taught by Euler in his textbook *Elements of algebra*.

Rhetorical Does Not Mean That There Is No Algebraic Language

Nesselmann's characterization of the rhetorical stage may lead one to think that if everything is written in words there is no algebraic language. Al-Khwārizmī's text is rhetorical because the algebraic calculations are expressed fully and in detail using Arabic words. However, to say that the algebra is then written in a natural language is to only tell half the story: al-Khwārizmī only uses Arabic words of course but besides setting technical meanings of the words, he schematizes and stereotypes the text. According to Anbouba, a phrase like *illā shay' fī illā shay' māl zā'id* (literally, "less thing by less thing, additive treasure"), which al-Khwārizmī writes in his book, "goes against grammar" (Anbouba, 1978, p. 72). Al-Khwārizmī breaks from the Arabic grammar as part of the elaboration of his algebraic language, which is done with no words other than Arabic. The new grammar produces schematic and stereotyped texts: "less thing" is more compact and manageable in a schematic and stereotyped text than the grammatically correct "the thing subtracted", and once it has been allowed to break the rules of grammar it can become independent of the complete expression that gave it meaning ("ten less thing", where the thing is subtracted from something), to designate something new ("less thing") with which algebraic calculations are made.

The words are from the Arabic language, but meanings and grammar have been modified to build an algebraic language into which problems are translated to solve them through operations that make sense in that language.

Syncopation can be seen as one more step in this process of schematization and stereotyping, with the examples that we have seen of abbreviations that end up not looking like such, as another step that separates the algebraic language from its origin in natural language.

Naming with a Number Instead of a Word

Chuquet proffered one of the main ideas that contributed to the elaboration of the symbolic sign system of algebra in the 15th century, but it went unnoticed and was later given again by Bombelli a century later. Both had the idea of not using the names of the species of numbers in the algebraic expressions and instead using their order number in the series of the species of numbers in continued proportion.

The idea of numbering the species of numbers in continued proportion can be traced back at least to as-Samaw'al's *Al-Bāhir en algèbre*, written in the 12th century (Ahmad & Rashed, 1972), and it appears also in Europe in the first half of the 16th century in Rudolff's *Die Coss*, Stifel's *Arithmetica integra*, Le Peletier's *L'algèbre* or Aurel's *Arithmetica algebratica*, to name a few. In these books, the numbers associated with the species of numbers were used as an auxiliary device to calculate the product (and the division) of two species of numbers, an operation that cannot be performed directly on the algebraic expressions of a system of signs in which the species of numbers are represented by their name or by an abbreviation.

Chuquet does not number the species of numbers previously expressed with their name, but when he defines the species of numbers he names them using the numbers, 0, 1, 2, and so on, which he calls his "denomination" (Marre, 1884, p. 152).

The above equation would be written in Chuquet's sign system as:

$$4^2 \; \tilde{p} \; 9^0 \; \tilde{m} \; 4^1 \; egaulx \; a \; 1^1$$

With slight differences, Bombelli would write[10]:

$$4^2 \; \mathrm{p} \; 9 \; m \; 4^1 \; Eguale \; à \; 1^1$$

Both equations are syncopated, but having a number as the name of the species of numbers makes it possible to calculate directly on the expression level following rules, a characteristic Nesselman ascribes to the symbolic stage.

A comparison of the corresponding terms in Bombelli's equation and in the same equation written in our symbolic sign system

$$4x^2 + 9 - 4x = x$$

shows that there is no sign in Bombelli for the unknown quantity, our x.

Symbolic Algebraic Expressions as Icons

It seems reasonable to think that the signs used in the algebraic language that Nesselmann called symbolic are symbols, that is, signs in which the relation with their object is conventional. However, Nesselmann's characterization of the symbolic language of algebra is not based on the nature of the individual signs: in his rhetoric stage all individual signs are words of Arabic, Latin, Greek, German languages, and the words of all these natural languages are conventional signs. That is, they are symbols and so an algebraic language made exclusively with symbols is not symbolic in Nesselmann's sense.

Actually, most of the signs of the symbolic language of algebra are conventional, but sometimes the author of the convention wanted the qualities of the sign to resemble those of his object, that is, that the newborn sign be an icon, and not a symbol. This is the case of the equal sign =.

Indeed, in *The Whetstone of Witte*, Robert Recorde introduced the sign by saying:

> And to avoide the tediouse repetition of these woordes: is equalle to: I will sette as I doe often in woorke use, a paire of paralleles, or Gemoine lines of one lengthe, thus ======, bicause noe 2 thynges, can be moare equalle. (Recorde, 1557, fo Ff1v)

He chose this sign for equality because no two things can be more equal than a pair of parallels. An icon of an abstract concept such as equality could not be a picture, but something that has equality as a salient property. The same happens with Pierre Hérigone's proposal in 1634 of 2|2 as the sign of equality in his *Cours mathématique*: 2|2 is an example of an equality, something that has the property of equality.

Individual signs can be icons rather than symbols, at least to the extent intended by their author. More interesting is the statement by Charles S. Peirce that "an algebraic formula is an icon", that he explains by stating that what makes it an icon is "the rules of commutation, association, and distribution of the symbols". In other words, the global structure of an algebraic expression is what makes it an icon. Peirce does not mention what the object it resembles is, of which object an algebraic expression is an icon. We would say that it stands for a network of quantities and relations, whose structure it resembles. Peirce adds that the importance of realizing that an algebraic expression is an icon is that

> a great distinguishing property of the icon is that by direct observation of it other truths concerning its object can be discovered than those which suffice to determine its construction. [...] This capacity of revealing unexpected truth is precisely that wherein the utility of algebraic formulae consists, so that the iconic character is the prevailing one.
>
> (Peirce, 1931–58, Vol. II, p. 158)

Ultimately, it is this iconic character of the expressions of algebra's symbolic sign system that inseparably binds algebra to the idea of structure.

Notes

1 The monetary context that corresponds to all the examples mentioned in the introduction is not, however, the only context of the problems that appear in the book. Now, the monetary context determines the terminology adopted for the objects of algebra, as we will see below.

2 These two technical terms appear in the title of the book and are the fundamental operations of algebraic calculus because they transform any equation into one of the canonical forms. The first removes all negative terms: *al-jabr* literally means "restoration" because it restores what is lacking. With the second, the equation is transformed into an equivalent one with only one term of each species of numbers.

3 My translation from Rashed bilingual (French and Arabic) edition, using also Hughes' edition of Gererdo de Cremona Latin translation. The Arabic word *māl* means "wealth" or "treasure", it does not mean "square". Gerardo de Cremona translated *māl* into Latin by *census* that also has the meaning of a quantity of money. From Cremona's *census* come the words used in the Renaissance in Europe, *censo* in Italian and Spanish, *çans* in French, *zensus* in German.

4 We sketch the steps towards modern algebra that produced the ideas and works of Lagrange, Abel, and Galois in Filloy, Rojano & Puig (2008, pp. 79–81).

5 Except for the modifications that have been made in digital media, we are not going to discuss here.

6 I am transliterating Greek words and letters. On the special case of the abbreviations of *arithmos* y *leipsis* (less), see Cajori (1993).

7 What Charles S. Peirce called a symbol.

8 In Cajori's book the sign for the thing appears here.

9 His new algebra is *logistica especiosa*, calculus with species, differentiated from *logistica numerosa*, calculus with numbers, for which Vieta used a syncopated sign system (see Oaks, 2018, on this).

10 In Bombelli's original, the raised numbers that represent the species of numbers are placed on horizontal arcs that I have not been able to include with the available word processor.

References

Abdeljaouad, M. 2002. Le manuscrit mathématique de Jerba: Une pratique des symboles algébriques maghrébins en pleine maturité. *Quaderni de Ricerca in Didattica del G.R.I.M. 11*, 110–173. http://math.unipa.it/~grim/MahdiAbdjQuad11.pdf.

Ahmad, S., & Rashed, R. (1972). *Al-bāhir en algèbre d'as-Samaw'al*. Damascus: Imp. de l'Université de Damas.

Anbouba, A. (1978). L'algèbre arabe aux IXe et Xe siècles. Aperçu général. *Journal for the History of Arabic Science, 2*, 66–100.

Cajori, F. (1993). *A history of mathematical notations*. Two volumes bound as one. New York: Dover.

Corry, L. (2007). From algebra (1895) to moderne algebra (1930): Changing conceptions of a discipline. A guided tour using the Jahrbuch über die Fortschritte der Mathematik. In J. Gray, & K. H. Parshall (Eds.), *Episodes in the history of modern algebra* (pp. 80–105). Providence: American Mathematical Society/London: Mathematical Society.

Filloy, E., Rojano, T., & Puig, L. (2008). *Educational algebra. A theoretical and empirical approach*. New York: Springer.

Giusti, E. (Ed.). (2020). *Leonardi Bigolli Pisano vulgo Fibonacci Liber Abbaci*. Firenze: Leo S. Olschki.

Høyrup, J. (2002). *Lengths, widths, surfaces a portrait of old babylonian algebra and its kin*. New York: Springer.

Hughes, B. (1986). Gerard of Cremona's translation of al-Khwārizmī's al-jabr: A critical edition. *Mediaeval Studies, 48*, 211–263.

Mahoney, M. S. (1971). Babylonian algebra: Form vs. content. *Studies in History and Philosophy of Science, 1*, 369–380.

Marre, A. (1884). *Le Triparty en la science des nombres par Maistre Nicolas Chuquet parisien*. Rome: Imprimerie des Sciences Mathématiques et Physiques.

Nesselman, G. H. F. (1842). *Versuch einer kritischen Geschichte der Algebra, 1. Teil. Die Algebra der Griechen*. Berlin: G. Reimer.

Oaks, J. (2012). Algebraic symbolism in medieval Arabic algebra. *Philosophica, 87*, 27–83.

Oaks, J. (2018). François Viète's revolution in algebra. *Archive for History of Exact Sciences, 72*, 245–302. doi: 10.1007/s00407-018-0208-0

Peirce, C. S. (1931–58). *Collected papers of Charles Sanders Peirce. Edited by Charles Hartshorne and Paul Weiss (vols. 1–6) and by Arthur Burks (vols. 7–8)*. Cambridge, MA: The Belknap Press of Harvard University Press.

Pérez de Moya, J. (1562). *Arithmetica practica y speculativa*. Salamanca: Matias Gast.

Puig, L., & Navarro, M. T. (2010). Protoálgebra en Babilonia (2a entrega). Métodos de solución. *Suma, 64*, 97–104.

Puig, L., & Rojano, T. (2004). The history of algebra in mathematics education. In K. Stacey, H. Chick, & M. Kendal (Eds.), *The future of the teaching and learning of algebra: The 12th ICMI study* (pp. 189–224). Boston/Dordrecht/New York/London: Kluwer Academic Publishers.

Rashed, R. (Ed.). (2007). *Al-Khwārizmī. Le commencement de l'algèbre*. Paris: Librairie Scientifique et Technique Albert Blanchard.

Recorde, R. (1557). *The whetstone of witte*. London: Jhon Kyngstone.

Serret, J.-A. (1877). *Cours d'algèbre supérieure. 4e édition*. Paris: Gauthier-Villars.

Souissi, M. (2001). *Feuilles d'automne ou En souvenir des Congrès et Colloques du patrimoine scientifique Maghrébo-arabe*. Beyrouth: Dar al-Gharb al-Islami.

Vieta, F. (1591). *In artem analyticem isagoge*. Turonis: Iametium Mettayer Typographum Regium.

3 Structure Sense at Early Ages

The Case of Equivalence of Numerical Expressions and Equalities

Carolyn Kieran and Cesar Martínez-Hernández[1]

*Students' experiences in learning arithmetic only
rarely foster an appreciation of structure
(Arcavi et al., 2017, p. 53)*

Introduction

In the numerical-algebraic world of school mathematics, equivalence has two faces: one, computational; and the other, structural. While the computational face tends to predominate at primary school, the structural is at the heart of secondary school algebra. Equivalence transformations based on structural properties are central to working with algebraic expressions and equations. Maintaining both the top-down and left-right equivalence of successively transformed equations in the equation-solving chain depends on students' algebra structure sense. While research that has been carried out with younger primary students on their interpretations of the equals sign has succeeded in enlarging their thinking about this symbol from operational towards more relational perspectives, other studies with secondary school students have pointed to their weakness in structuring ability and their difficulty in drawing upon underlying properties to form equivalent structures. More recent research on what has been referred to as early algebraic thinking has shown promise in developing aspects of students' structural thinking at the primary school level; however, much of this work has concentrated on generalizing figurative patterns and only indirectly on structure within this activity. The multifaceted components of equivalence within a numerical setting remain largely unstudied – despite their relevance for later work in school algebra and the importance of beginning to develop such algebraic thinking at the primary levels of schooling. The present study aimed at partially filling this gap by fostering the growth of 10- to 12-year-old Mexican students' structure sense regarding numerical expressions and equalities. The core of the study included tasks and group interviews that focused

DOI: 10.4324/9781003197867-3

on indicating and justifying the truth-value of numerical equalities within the activity of generating equivalent equalities – where equivalence was grounded in properties related to the order and addition structures, properties such as decomposition, as well as the reflexive, symmetric, and transitive properties of equality. The results of the study illustrate the role played by decomposition in the students' evolution from a computational to a structural perspective on equivalence. The discussion highlights the nature of the interviewer's interventions that fostered this evolution.

This chapter is structured as follows: The initial two sections, which are more theoretical in nature, deal in the first case with mathematical equivalence and in the second with structure, structuring, and structure sense. The next section reviews the relevant research literature related to the main theoretical constructs of our study and thereby motivates its empirical underpinnings. The subsequent section describes the methodological components of the study and, in particular, elaborates on the way in which the tasks were designed. The following two sections present the results of the study and their discussion. The chapter ends with a brief concluding section. (Note that, throughout, our use of the term *equality* refers to those mathematical objects that contain an equals sign and numerical terms, but no variable or unknown terms; we reserve the term *equation* for those cases when variables or unknowns are present.)

On Mathematical Equivalence

While the actual term *equivalence* and its definition came into use as late as the 20th century, Asghari (2019) recounts the history of this notion as a lived experience that goes back to much earlier times. For example, from Euclid we have that "straight lines parallel to the same straight line are also parallel to one another" and later from Hilbert, "if two segments are congruent to a third one, they are congruent to each other". The notion of equivalence, even if not called equivalence, has clearly been and remains a fundamental aspect of mathematics.

Eventually, mathematical equivalence arrived at its modern definition: An equivalence relation is a binary relation that is reflexive, symmetric, and transitive. The relation "is equal to" is the canonical example of an equivalence relation, where for any objects a, b, and c:

$a = a$ (reflexive property);

if $a = b$, then $b = a$ (symmetric property); and

if $a = b$ and $b = c$, then $a = c$ (transitive property).

Equality is a relationship between two quantities, or more generally two mathematical expressions, asserting either that the quantities have the same value, or that the expressions represent the same mathematical object, or that an object is being defined – as in $\pi = 3.14159....$

With respect to the numerical world, equivalence undeniably has a computational dimension. This dimension arises from the above definition of the equivalence relation "is equal to" whereby the two quantities "have the same value" – as in $5 + 9 + 3 = 10 + 7$ is true because both sides evaluate to 17. But, clearly equivalence also has a structural dimension. This latter dimension arises from the same definition of equivalence, but capitalizes on the reflexive property where $a = a$ with a being a mathematical expression – as in $5 + 9 + 3 = 5 + 9 + 3$ where the structure of the non-computed sum of each side of the initial equality ($5 + 9 + 3 = 10 + 7$) is reflected in the same decomposition on the two sides of the second equality (i.e., the 10 has been decomposed into $5 + 5$ and the 7 into $4 + 3$, which with a recomposition of the $5 + 4$, along with the use of the associative property of addition, leads to the right-hand expression $5 + 9 + 3$ of the second equality) and whereby the truth-value of the equality statement is visibly obvious without computing. The role of properties is indeed especially prominent within the structural dimension of equivalence.

On Structure, Structuring, and Structure Sense

As is the case with equivalence, structure is another of the fundamental ideas of mathematics. However, unlike *equivalence,* which has been well defined, *structure* is not. It is, however, often treated within the mathematics education community as if it were an undefined term (Kieran, 2018). While many researchers use the term *structure*, it is just assumed that there is universal agreement on its meaning (Mason et al., 2010). In fact, Venkat et al. (2019) argue that the community seems to have difficulty in defining *structure* in a coherent way.

For Mason et al. (2009), the structural is closely intertwined both with the general and with attending to properties. We would argue, however, that the intertwining of the structural with the general has resulted in giving more emphasis to generalizing within arithmetic and algebra, and much less to the structural. In fact, the process of seeing structure has to a large extent been neglected by generalization-oriented activity in that the structural part of generalizing has been considered merely as inherent to the process of generalizing. There is no question that high school algebra involves working with generalized forms, such as equations and expressions, but this activity also requires the ability to see structure in these forms and to re-express these already generalized forms with equivalent structures. Hoch and Dreyfus (2004) stress precisely this aspect when they describe *structure sense* as consisting of the following:

> The [abilities] to see an algebraic expression or sentence as an entity, recognize an algebraic expression or sentence as a previously met structure, divide an entity into sub-structures, recognize mutual connections between structures, recognize which manipulations it is possible to perform, and recognize which manipulations it is useful to perform. (p. 51)

However, they found such structuring ability to be weak in the high schoolers they studied – even more so, the students' ability to draw upon underlying properties to form equivalent structures.

Similarly, Linchevski and Livneh (1999), who were the first ones to use the phrase *structure sense,* call attention to young students' difficulties with using knowledge of arithmetic structures at the early stages of learning algebra. They thereupon suggest that instruction in arithmetic be designed to foster the development of structure sense by providing experience with equivalent structures of numerical expressions and with their decomposition and recomposition. However, lest it be thought that decomposing and recomposing relate primarily to the numeric and less so to the algebraic world, such is not the case. For example, even the first steps of the solving for x of a very simple equation, such as,

$$3x^2 - 9x - 30 = 0$$

$$(3x+6)(x-5) = 0$$

$$3x+6 = 0 \ or \ x-5 = 0 \ ...,$$

can involve decomposing an expression of the given equation into an equivalent structure – in this case, $3x^2 - 9x - 30$ into $(3x + 6)(x - 5)$. This decomposing, which relies on students' algebra structure sense so as to maintain the top-down equivalence between the initial equation and its transformed version in the second line, draws upon the distributive property. Of course, other properties are also involved in maintaining the equivalences of the equation-solving chain, as for example the zero-product property that links the second line with the third.

Let us look at a slightly different type of algebraic object. This next three-line example involves showing that a given equation is true for all values of x:

$$(x-3)(4x-3) = (-x+3)^2 + x(3x-9)$$

$$4x^2 - 3x - 12x + 9 = x^2 - 3x - 3x + 9 + 3x^2 - 9x$$

$$4x^2 - 15x + 9 = 4x^2 - 15x + 9.$$

For this particular example, recomposing the expressions of both sides of the given equation by means of distributivity yields the equivalent second equation, while another recomposing in accordance with the addition structure produces the third equation. With this example, we note how the application of properties not only preserves the rather invisible top–down equivalence between each equation and its successor, but at the same time produces a quite visible left–right equivalence of both sides in the third line – the left–right equivalence having been invisible until the third line. With this now explicit sameness of the left and right sides, the equation is clearly seen to be true for all values of the variable. The equivalence considerations that are illustrated in these two examples – equivalences that are tied to the use of decomposition and recomposition – are central components of algebra structure sense, components that we would argue can be fostered at the primary school level within a numerical context. But, first, we need to say more about structure and structuring.

In an early effort to deconstruct the notion of structure, Kieran (1989) distinguished between the surface structure and the systemic structure of numeric and algebraic expressions and equalities/equations. While the surface structure was viewed as being related to arrangement or disposition, the systemic structure was characterized in terms of the basic field properties of arithmetic – a perspective that we now believe to be too narrow for fostering the development of structure sense in primary school students. For a broader perspective on structure and properties, we turn to Freudenthal.

Freudenthal (1991) points out that the system of whole numbers constitutes an *order structure* where addition can be derived from the order in the structure, such that for each pair of numbers a third, its sum, can be assigned. The relations of this system are of the form $a + b = c$, which he refers to as an *addition structure*. In his book *Didactical Phenomenology of Mathematical Structures*, Freudenthal (1983, pp. 112–113) goes on to describe the *multiplicative structure* of the natural numbers in terms that comprise more than the act of multiplying. It is the whole complex of relations $a \times b = c$, possibly also expressed as $c / b = a$, and complemented by $a \times b \times c = d$, $a \times b = d / c$, and all other relations one would like to consider in this context. It encompasses such properties as commutativity, associativity, distributivity, equivalence of $a \times b = c$ and $c / b = a$, and many more properties of this kind. According to Freudenthal, the structure of the natural numbers also allows for prescribing c in the relation $a \times b = c$ and asking for its splittings into two factors. Freudenthal asserts further that c can be split into its prime factors, with divisors and multiples being other means of structuring. He adds as well that tying the order structure to the multiplicative structure yields the property that, given the product, increasing one factor means decreasing the other. His use of the term *means of structuring* is important for it not only admits the learner into the picture, but also, as he emphasizes, these alternative structurings allow for a better "grasp of the structure of **N**" and its multiple properties.

In line with Freudenthal's position that there are multiple means of structuring, all of which entail a multitude of properties – properties that can be characterized in a manner that is not restricted to their succinct axiomatic formulation within the basic properties of arithmetic – the discussion of properties related to structural thinking with numbers and numerical operations can be broadened in a way that extends beyond the usual manner in which basic properties are cited – in short, all sorts of properties related to the order, addition, and multiplicative structures, as well as the properties of the equivalence relation.

The above theoretical literature on equivalence, structure, and multiple-property structurings allows us to open up the idea of equivalence of numerical expressions and equalities at the primary school level. First, the discussion on equivalence leads us to characterize *structure sense in the numerical domain* as engaging in thinking structurally about and representing number and numerical operations in such a way that decisions about equivalence can be made without resorting to computing, but instead structurally – by relating to the structure of number and its properties. Second, the broadening of perspective on structure and structuring leads us to consider decomposing, composing, and recomposing also as properties – properties that are tied to the order, addition, and multiplicative structures. The use of these properties is also anchored in the symmetric property of equality; that is, symmetry allows for the rewriting of the addition fact of $5 + 7 = 12$ as $12 = 5 + 7$. However, these addition facts can also be interpreted and acted upon as 5 plus 7 being composed into 12 and as 12 being decomposed into 5 plus 7 – in fact, any combination of its various addends – which seems a more dynamic and appropriate approach than that of invoking symmetry and one that we hypothesize may be at the heart of students' structuring activity in primary school.

On Prior Research in This Area

From the 1970s, when research on the ways in which students view the equals sign began to emerge, and continuing on up to the present day, studies have been carried out on the understandings that students come to have regarding the equals sign, as well as on the ways in which instruction can affect those understandings (e.g., Carpenter et al., 2003; Kieran, 1981; Knuth et al., 2006; Lee & Pang, 2021). In brief, young students' views of the equals sign have been characterized in terms of varying degrees of "operator" (i.e., as a "do something signal") and "relational" thinking (i.e., as a "sign denoting the relation between two equal quantities") – these varying views usually considered a reflection of the ways in which students have been taught (e.g., Li et al., 2008). The kinds of tasks that have generally been used in the studies gauging students' views of the equals sign have tended to include (i) true-false equalities where students are asked to state their truth-value, (ii) open sentences requiring them to determine which number will make the sentence true, and (iii) the request to provide a definition of the equals sign. By means of such tasks, Rittle-Johnson et al. (2010), for example, were able to assess how U.S. students' knowledge of mathematical equivalence – *equivalence* being the term chosen for use by these researchers – develops through the 7- to 12-year-old age range. Their results suggested that by about the 5th grade most students have begun to compare both sides of an equality or equation and thus hold a basic relational view of equivalence (e.g., they can accept equations and equalities with operations on both sides).

Very little of the existing research literature on the equals sign and equivalence at the primary school level has included explicit attention to decomposing, composing, and recomposing. While Jacobs et al. (2007) noted that students recomposed the expressions 5×499 and $1488 + 375 - 373$ into related ones (i.e., $5 \times 500 - 5$ and $1488 + 2$), these researchers emphasized that the students did so in order to simplify calculations and not to indicate equivalence. Similarly, in Carpenter et al. (2003), students were able to rewrite expressions such as $10 + 0$ as $100 - 90 + 10 - 10$ and then express the two as $10 + 0 = 100 - 90 + 10 - 10$, but this was done in the context of re-expressing zero in terms of complementary operations. In another study, Molina and Ambrose (2006, p. 113) asked students to write true sentences of the form "$\_ + \_ = \_ + \_$", to which one student responded with $90 + 200 = 200 + 10 + 10 + 20 + 30 + 20$, expressing commutativity imaginatively in a way that involved a "string of addends". Students' limited experience with the various facets of decomposing within activity related to explicitly generating equivalences was also revealed in a study by Warren (2003), who found that only 5% of the 672 students of 7th and 8th grades she tested responded that there was an unlimited number of possibilities for answering the question: "Write other sums that add to 23; how many can you write?"; and fully one-third of the students failed to respond to the question in any way.

While research with a structural orientation has been scattered and quite rare up to now at the primary school level (Venkat et al., 2018), even rarer are studies on number, numerical operations, and equality/equivalence with explicit attention to underlying properties (Kieran, 2018). Even though Linchevski and Livneh advocated for this kind of structuring activity within arithmetic more than two decades ago, the need for learning to restructure an arithmetic expression into an equivalent one continues to this day to be an area in need of research attention (Asghari & Khosroshahi, 2017). Little if any research has focused on the kind of structural thinking that is involved in the generation of equivalent numerical equalities by younger age students. This largely untouched area of research is one that we believe to be essential to the development of students' early algebraic thinking and their later algebra structure sense (for those wishing to read more on various characterizations of early algebraic thinking, see, e.g., Blanton et al., 2018, p. 30; Kaput, 2008, p. 11; Kieran, 2007, p. 714; Radford, 2018, p. 8).

On the Research Study: Methodology

Background Remarks

Our goal in this study was to foster the development of young students' structure sense by providing them with experiences requiring the generation of numerical equalities equivalent to given equalities. The main question in our research was oriented towards investigating the ways in which students determine and re-express the truth-value of the expressions of the given equalities within tasks aimed at constructing equalities equivalent to the given ones. We were interested in studying the evolution of their strategies in the face of specific tasks and group interactions involving interviewer and students. The study took place during two time periods. The first period, Part 1, occurred during the middle of the school year and involved six students; the second period, Part 2, took place at the end of the school year and involved three of the original six students – the three who had been the most verbal of the six and who had tended to participate more fully during the earlier interview sessions.

Participants and Their Prior Experience in This Area

Six students from the 6th grade, ages 10 and 11 years, from a public community school of a small Mexican municipality participated in the first part of the study, which was conducted when the students were halfway through their last year of primary school (December 2017). The 2nd part of the study, in which three of the six original students continued to participate (by which time they were 11 and 12 years of age), was conducted in June 2018 at the moment when the students were just finishing their last year.

This grade level was chosen because such students are on the verge of completing their primary school education and have been exposed to the official Mexican public education curriculum. In the curriculum for the elementary school (Secretaría de Educación Pública (SEP), 2016), the equivalence of numerical expressions is not mentioned. Moreover, a survey that we conducted of the grades 5 and 6 textbooks used by the students of our study found no tasks that were specifically related to developing structural views of equalities. Additionally, prior to the unfolding of the designed activities, the classroom teacher of these students reviewed the tasks. In her opinion, the students had never solved similar tasks; they had only worked with the use of the equals sign in a computational manner. Thus, the students had had no prior experience in thinking structurally about equalities or, in particular, with generating equivalent numerical equalities. They had however seen and worked with equalities containing at least two numerical terms on each side of the equals sign.

Method

The data collection technique was that of the Group Interview, a method that involved the students first working individually on a given task question or set of questions. For each subset of questions, the individual work was followed by an interviewer-orchestrated, discussion segment where the students would share their responses with the rest of the group. During this group sharing, the interviewer might probe their thinking by asking for clarification or further elaboration; he might also pose questions that could lead the students to consider alternatives or nudge their thinking into a different direction. The group interview was thereby designed to offer the students ample opportunity to verbalize their reasoning and to build upon the ideas of their fellow participants.

Data from Part 1 of the study were obtained during three sessions, one session per whole task-set, with sessions lasting about 60 minutes each. The Group Interviews took place in one of the rooms of the school. All six students participated in each of the three sessions. Data from Part 2 of the study were obtained during one session that lasted about 60 minutes and involved three of the original six students. The data sources for both parts of the study include the individual students' worksheets, video-taped footage of the interviewer interacting with the group of students and the recording of all their verbalizations during each of the sessions, and the researcher's field notes. All interactions and task-sets were in the Spanish language. The interviewer was the second author of this chapter.

Design of the Tasks

For Part 1 of the study, three task-sets were designed in order to explore students' strategies related to structure sense regarding equivalence of expressions and equalities. The first two of these task-sets did not include the equals sign; they were designed to uncover by means of different formats the various ways in which students might be thinking structurally about equality relationships. Task-set 1 aimed at identifying the ways in which students relate two numbers a and b with a third one c (i.e., its sum) and the rationale they use (e.g., "Can the number 7 be written from the numbers 6 and 1? If so, how?"). Task-set 2 was based on a sequence proposed by Schifter (2018), that is, "14 + 1, 13 + 2, 12 + 3, 11 + 4, 10 + 5", and inquired into the ways in which students would describe the regularity they were seeing, as well as how they might transform, for example, 13 + 2 into 14 + 1. Task-set 3 involved the use of the equals sign in numeric equality statements and included two or more numerical terms on each side of the equals sign (i.e., $4 + 5 = 4 + 3 + 2$, $480 + 6 + 123 = 486 + 123$, $172 + 10 + 75 = 182 + 50 + 25$, $150 - 70 = 125 + 25 - 70$, and the non-equality $2 + 8 = 1 + 1 + 5$). The design of the task-set consisted of two types of questions: one asking students whether a given equality statement was true and to explain why, and the other to rewrite the true equalities in another way so as to show they were true. The objective was to study if and how students decompose and recompose the numbers of such equalities in order to indicate their truth-value.

Due to the computational nature of the results obtained for Task-set 3 in Part 1 of the study, a new task-set (Task-set 4) was designed and served as the main focus for the 2nd Part of the study. Task-set 4 involved equalities of equivalent numerical expressions (i.e., $10 + 7 = 5 + 12$, $530 + 200 = 300 + 430$, and $8 + 2 + 16 = 10 + 12 + 4$). However, the task instructions specifically requested that the students <u>not</u> calculate the total of each side in order to show that the equality was true. For each equality that was proposed, the task was structured as follows:

- A true numerical equality is proposed to the students;
- Students are to show that the numerical equality is true, but with the explicit request that they not calculate the total of each side in order to show its truth-value;
- The students are asked for an explanation of their reasoning.

In line with the broader approach to structure and structuring that was presented earlier, the structural properties that were considered to be at play in students' generation of equivalent numerical expressions and equalities were those related in particular to decomposition, composition, and recomposition. In accordance with the addition structure $a + b = c$ (where a, b, and c are whole numbers), a and b – by means of the *property of composition* – are related by equality to their sum c. The equality $a + b = c$ is equivalent by the *property of symmetry* to $c = a + b$. This latter representation involving the reversed form of the addition structure can be extended; that is, by the *property of decomposition*, the number c can be expressed not only as the sum of these two addends, but also as the sum of any of its other possible addends (e.g., $c = d + e + f$). Note that the number of terms in these various decomposed expressions is not fixed – that is, they can be two or more. Other properties that would be called upon in the given tasks include the common-form property of equivalence (i.e., for a pair of expressions, converting either/both expressions into a common form can be used to indicate equivalence) and, perhaps more implicitly, the transitive property of equality.

On the Research Study: Results

The development of students' sense of structure with respect to the generation of equivalent numerical equalities was found to evolve through five specific phases. Data providing evidence of these five phases are highlighted below and are consolidated from partial findings presented in conference research reports by Martínez-Hernández and Kieran (2018, 2019, 2020a, 2020b; also Kieran & Martínez-Hernández, 2021). Note that the three students (S1, S2, and S3 – two girls and one boy, respectively) who are featured in the excerpts below are the same three whose work is reported in those conference papers.

The Phase of Computation Without Decomposition

Students' sense of structure was initially dominated by a computational perspective that was evident right from the beginning of Part 1 of the study. We briefly summarize the approaches used by the students for Task-sets 1 and 2, before moving into the results involving the tasks with an equals sign.

Task-set 1: This task-set allowed us to see whether students use the equals sign spontaneously and, if so, in which way. Three of the items were the following:

1   *Can the number 7 be written from the numbers 6 and 1? If so, how?*
2   *Can the number 19 be written from the numbers 14 and 5? If so, how?*
3   *Is it correct to write the number 7 as 3 + 4? As 8 + 2? Explain.*

The students answered affirmatively for items 1 and 2, and explained their thinking by means of a computation involving the property of composing the addends. For example, S1 related 7 with 6 and 1 by writing $6 + 1 = 7$ for item 1. None of the students wrote, for example, $7 = 6 + 1$, which would have reflected the use of the symmetric property and a view of number as being decomposable into its parts. In item 3, S3 answered in the same computational way, stressing *the result that must be obtained*.

The data from Task-set 2 (involving the sequence $14 + 1$, $13 + 2$, $12 + 3$, $11 + 4$, $10 + 5$) revealed the features that students observed of the $a + b$ form of expressions, as well as the possibility of their transforming one expression of the sequence into another expression of the same sequence. Regarding the first aspect, students' answers showed that they identified the regularity in the sequence. For instance, they wrote about *the involved operation (addition)*, *the sum (the result)*, and *the order* in the sequence (*increase* and *decrease* of the addends).

The answers that students produced regarding the presented sequence relate, on the one hand, to the kind of thinking they showed throughout Task-set 1. That is, they identified the expression as an operation that must be carried out in order *to obtain a result*. This feature was clear in S2's response when she wrote: *"…all of them are additions, and they are not answered and all the additions result in 15."* On the other hand,

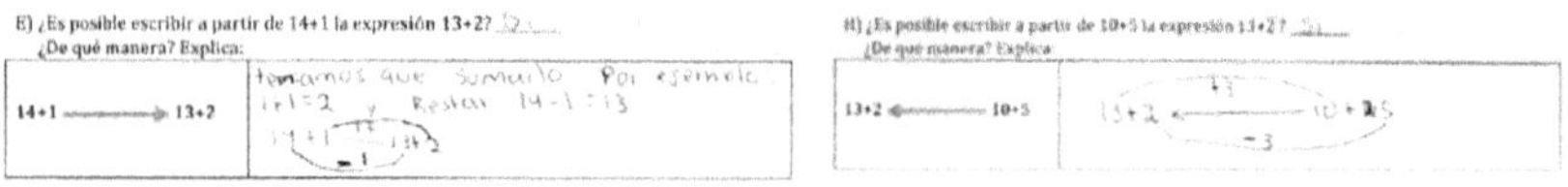

[E] Is it possible to write from 14+1 the expression 13+2?
Yes
In which way? Explain:
We have to add. For instance, 1+1=2 and subtract 14 -1=13]

[H] Is it possible to write from 10+5 the expression 13+2?
Yes
In which way? Explain:
(Student illustrates an additive compensation strategy)]

Figure 3.1 S1's additive compensation strategy.

there is some evidence of structural thinking related to pattern recognition, seen in S3's work when he stated: *"… and the bigger number becomes small, the smaller becomes big."* However, he did not relate the feature he observed to the equivalence of the expressions in the form of an equality.

In the second part of Task-set 2, the students were asked how to obtain one expression in the sequence (e.g., 13 + 2) from another (e.g., 14 + 1). The students used diagrammatic means to show how an addition was compensated for by a corresponding subtraction (see S1's work in Figure 3.1; please note that the English translation that is provided in these figures was not part of the students' worksheets).

For Task-set 3, all of the items included the equals sign and involved, for some equalities, smaller numbers and for others, larger numbers – the latter case designed to investigate whether larger numbers might provoke different strategies, and perhaps deter the use of computational approaches. For the first type of question of Task-set 3 – asking students whether a given equality statement was true and to explain why – it was found that students could accept without hesitation equalities in the form $a + b = c + d$. However, they continued to justify the truth (or falsity) of each of the five given equality statements by calculating the result on each side – whether the equality involved small or large numbers (see Figure 3.2).

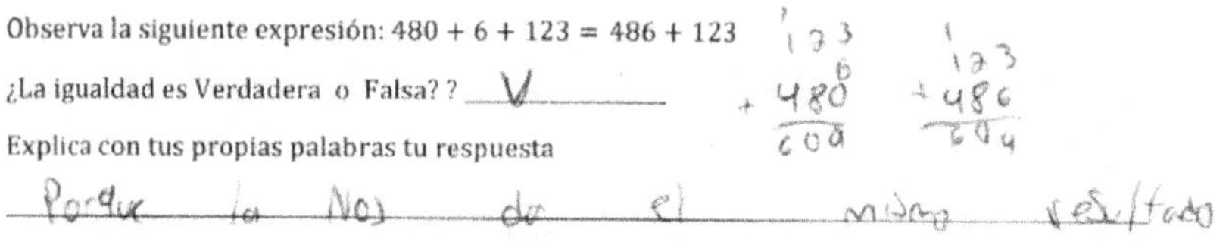

[Observe the following expression: 480 + 6 + 123 = 486 + 123
Is the equality True or False? _T_
Explain with your own words
        Because we get the same result]

Figure 3.2 S3's computational strategy (note S3's additions off to the side).

The Phase of Computation with Unrelated Decompositions of Both Sides

Each of the equality tasks in Task-set 3 also included a second type of question: "In which other way could you rewrite the given equality?" Even though the students had computed the totals for each side of the given equalities in order to verify their truth-value, they were able to generate alternative forms of each equality by means of two different, but related, decomposing strategies. In the first of these strategies, they decomposed each of the given addends of each side of the equality in a different way (see S1's work in Figure 3.3). The second, which was clearly informed by first calculating the *total for each side*, involved looking for two numbers for one side of the equality that would yield the already-computed total and another two different numbers for the other side (see S3's work in Figure 3.4). As can be observed from the students' responses to this question, there was no inclination for the students to re-express the equalities in such a way that both sides of the equalities looked alike. Each side of the equality was decomposed independently of the other side. As seen in Figure 3.3, S1 noted that the results (on both sides) were the same. Clearly, the students' view of the left–right equivalence of the expressions of a numerical equality was computational, even if they were beginning to use the property of decomposition.

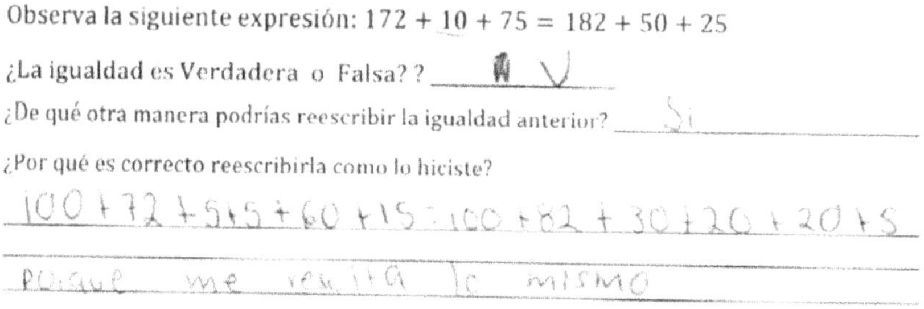

Observa la siguiente expresión: $172 + 10 + 75 = 182 + 50 + 25$

¿La igualdad es Verdadera o Falsa? ? ____ **A** ✓

¿De qué otra manera podrías reescribir la igualdad anterior? ____ Si

¿Por qué es correcto reescribirla como lo hiciste?

100 + 72 + 5 + 5 + 60 + 15 = 100 + 82 + 30 + 20 + 20 + 5

Porque me resulta lo mismo

[Observe the following: $172 + 10 + 75 = 182 + 50 + 25$
Is the equality True or False? _T_
In which other way could you re-write the given equality?
Yes.
Why it is correct re-write the expression in such a way?
$100 + 72 + 5 + 5 + 60 + 15 = 100 + 82 + 30 + 20 + 20 + 5$
Because I get the same result]

Figure 3.3 S1's equality rewriting.

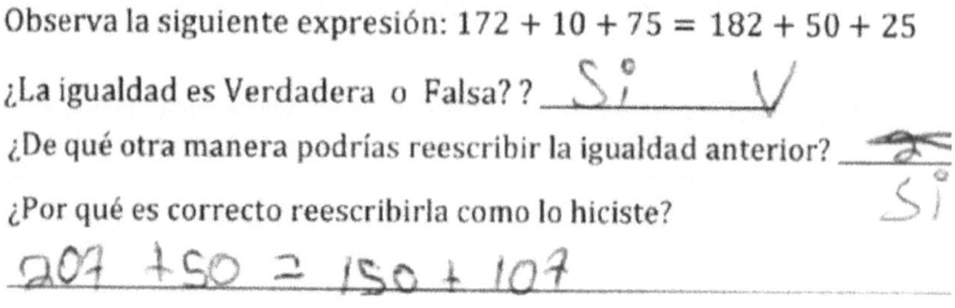

Observa la siguiente expresión: 172 + 10 + 75 = 182 + 50 + 25

¿La igualdad es Verdadera o Falsa?? _Sí_   ✓

¿De qué otra manera podrías reescribir la igualdad anterior? _____

¿Por qué es correcto reescribirla como lo hiciste?   Sí

207 + 50 = 150 + 107

[Observe the following: 172 + 10 + 75 = 182 + 50 +25
Is the equality True or False? *Yes T*
In which other way could you re-write the given equality?
Yes.
Why it is correct re-write the expression in such a way?
207 + 50 = 150 + 107]

Figure 3.4 S3's equality rewriting.

The Phase of Ad Hoc Decomposition of One Side and Copying to the Other Side

When the three students came together again for Part 2 of the study six months later, they were presented with Task-set 4 in which they were requested to show the truth-value of each of the given equalities, but this time with the explicit requirement that it be done without first calculating the total of each side of the equalities. There were initially looks of puzzlement among the students, as if to say, "What other way is there?" Despite the explicit request to not calculate, the students all used the computational approach of totalling each of the two sides to show that the first equality, $10 + 7 = 5 + 12$, was true. An example of this is S1's written answer where she states: "It is true because if we add the two sides, the sums that are the results are the same."

Because the students had no idea how to answer the question ("Without calculating the total of each side, show that the equality is true"), some interviewer-prompting within the group discussion became necessary (the English transcriptions below are verbatim translations of the original Spanish version; I is the Interviewer). As seen above, the students had already indicated during Part 1 of the study that they could decompose a given equality, after they had first computed the total of each side. The transition that was needed was for them to carry out a decomposition, but without first computing the total of the expressions on each side of the equality. The interviewer began his intervention by using the same form of questioning that had been used in the prior Task-set 3:

> I: Is there another way to rewrite this equality? [the Interviewer writes on the board the equality $10 + 7 = 5 + 12$]
> S2: As a subtraction
> S3: As a subtraction [inaudible]
> I: Alright, as a subtraction, according to S2 [...] For example, S2?
> S1: Twenty minus three [note that S1 interjects an answer]
> S2: Equals eighteen minus one [S2 completes the equality $20 - 3 = 18 - 1$].

As can be observed from this excerpt above, the strategy followed by S1 and S2 was to search for two numbers in such a way as to preserve the mentally-calculated total value for each side. They managed each side of the equality in an independent way, guided by the computed total. So the interviewer tried another approach:

> I: Would there be a way to write also [referring to rewriting the given equality] but using, say, these same numbers? [points to the board to the equality $10 + 7 = 5 + 12$]
> S3: Yes.
> I: [...] For instance, could this [points to the left side, $10 + 7$, but S3 interrupts]?
> S3: $5 + 5 + 5 + 2$ [S3 verbalizes the expression].
> I: Ok, S3 says that this [referring to the left side of the equality $10 + 7 = 5 + 12$] could be written as $5 + 5 + 5 + 2$ [writes on the blackboard the expression stated by S3]. Is this OK?
> S1 and S2: Yes [both at once].
> S3: Is equal to [inaudible].
> I: [...] This [referring to the right side of $10 + 7 = 5 + 12$], in which other way? Look, this [referring to $5 + 5 + 5 + 2$] already has a form of, I mean, it [$10 + 7$] can be re-expressed in this way [referring to $5 + 5 + 5 + 2$] [...]. Ok, can this [the right side $5 + 12$] be rewritten identically to this [referring to $5 + 5 + 5 + 2$]?
> S3: Yes.
> I: Why?
> S2: It gives the same.
> S1: Because it gives the same.
> I: How would you rewrite it?
> S3: $5 + 5 + 5 + 2$ [verbalizes the expression].
> I: Ok, $5 + 5 + 5 + 2$ [completes the equality $5 + 5 + 5 + 2 = 5 + 5 + 5 + 2$ as proposed by S3, on the blackboard].

As can be seen from the preceding verbatim, S3 was able – guided by the questioning put forth by the interviewer – to propose an equivalent form ($5 + 5 + 5 + 2 = 5 + 5 + 5 + 2$) for the equality $10 + 7 = 5 + 12$ in such a way that both sides of the equality had the same form. However, it was not yet clear whether or not the right side of the initial equality had actually been decomposed or the transformed expression simply recopied to the other side. As of this moment, S1 and S2 justified the second equality based on the preserved total. This is seen when S1 answered: *"Because it gives the same."* Nevertheless, the form proposed by S3 had the appearance of a decomposition of each side of the initial equality. This suggests that perhaps S3 was leaving behind the idea of looking for two or more numbers that preserve the total. The evidence for this shift is derived from the following verbatim, where both S2 and S3 indicated that their initial strategies were beginning to evolve:

> I: Does this 5 [Interviewer points to the first number 5 on the left side, $5 + 5 + 5 + 2$, of the equality $5 + 5 + 5 + 2 = 5 + 5 + 5 + 2$] come from any part here [points to the initial equality $10 + 7 = 5 + 12$]? [...] Or, are you looking for two or more numbers in order to get 17 [the total]?
>
> S3: Yes, I rewrote the two 5s [the first two numbers of $5 + 5 + 5 + 2$] from the 10 [referring to the 10 in the left-hand expression $10 + 7$], and from the 12.
>
> S2: From the 7! [S2 interrupts, referring to the 7 in the expression $10 + 7$].
>
> S3: From the 7, we had 2 left, that is why I wrote the plus 2 [S3 explains that the 7 is rewritten as $5 + 2$].

And when the interviewer asked about the rewritten right side, the following ensued:

> I: And, how would you justify these? [referring to the right side of the equality $5 + 5 + 5 + 2 = 5 + 5 + 5 + 2$]
>
> S2: The 5 comes from the 5 [of the right side of the initial equality $10 + 7 = 5 + 12$], the two 5s and the 2 come from the 12.

As suggested by the verbatim, S2 and S3 were now able to relate the terms of the decomposed expressions to the corresponding expressions of the initial form of the equality. That is, they explained, through decomposition, where each number came from in $5 + 5 + 5 + 2 = 5 + 5 + 5 + 2$ – an approach that is synthesized in Figure 3.5. However, it was still not obvious whether or not their justification was of an *ad hoc* nature, in order to have the expression correspond with the right side ($5 + 12$) of the initial equality. We return to this issue shortly.

Based on this new strategy of rewriting each side (or at least one side) through decomposition of numbers and getting both sides of the

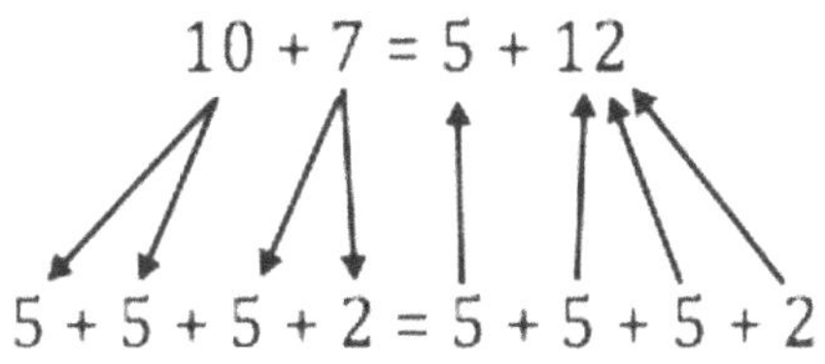

$$10 + 7 = 5 + 12$$

$$5 + 5 + 5 + 2 = 5 + 5 + 5 + 2$$

Figure 3.5 Equality rewritten by means of decomposition.

rewritten equality to have the same form, the interviewer returned to S3's equality to inquire into the need (or not) to calculate the total:

I: Seeing it in this way [points to the equality 5 + 5 + 5 + 2 = 5 + 5 + 5 + 2], is it necessary to add up in order to decide if the equality is true? Would you still add [calculate the total of each side] or is the addition no longer necessary?

S3: It is no longer necessary for me.

I: Why not, S3?

S3: Because it is easy to see what will be the result.

I: Ok [...] What is this and this expression like? [pointing to both sides of 5 + 5 + 5 + 2 = 5 + 5 + 5 + 2]

S2: The same.

S3: The same.

I: Then, is it necessary to add?

S1: Oh, no!

S2: No, because [inaudible]

S3: But to know the result of each one? [referring to the total of each side]

S2: No, but if it is the same [in the same form], obviously it will give the same [the total will be the same]. If the expression is the same, it will be equal, it will give the same.

As can be noted from the above verbatim extract, S3 related the decomposed equality to both sides of the initial equality, but still considered the total as a means to validate it. S2 and S1 supported the idea that the decomposed equality of S3 was all that was needed for validation, due to the common form in which both sides could be rewritten. In this regard, it is important to note how S2's speech and thinking had changed. At the beginning of the task, S2 stated that the total *"is" the same*, which suggests that S2 had calculated the total. But now S2 justified with another tense: *"will be" the same*, which means that S2 was aware that she would get the same total, but that it was not necessary to calculate it. It was enough to observe that both sides of the equality were written in the same form.

The interviewer subsequently asked the students to rewrite in another different form (different from $5 + 5 + 5 + 2 = 5 + 5 + 5 + 2$) the initial equality $10 + 7 = 5 + 12$, in such a way that both sides would look alike. However, S1 genuinely decomposed only the left side $(10 + 7)$ of the initial equality to obtain $2 + 2 + 2 + 2 + 2 + 2 + 2 + 3$ and then copied this expression onto the right side. This was made obvious because she could not relate the expression $2 + 2 + 2 + 2 + 2 + 2 + 2 + 3$ (i.e., the right side of the rewritten equality) with the right side $(5 + 12)$ of the initial equality when asked to do so. After a few futile attempts, S1 started anew. She proposed a different expression for the right side, $2 + 2 + 1 + 5 + 5 + 2$, and then rewrote the left side of the initial equality with the same expression, which she could justify.

Now, the students were clearly able to rewrite an equality by means of a decomposition that distinctly showed equivalence of the expressions on each side of the equality and where the truth-value of the equality was apparent from its rewritten form. This is seen, for example, in the explanation offered by S1: *"Because if I look and compare each number, they are the same and obviously it is the same"*.

The nature of their *ad hoc* decomposition and copying strategy was also made apparent in equalities involving larger numbers. For instance, for the equality $530 + 200 = 300 + 430$, S1 proposed: $100 + 200 + 100 + 100 + 30 + 100 + 100 = 100 + 200 + 100 + 100 + 30 + 100 + 100$; and S3 proposed: $300 + 200 + 30 + 100 + 100 = 300 + 200 + 30 + 100 + 100$. But, in each case, both students decomposed only the left side of $530 + 200 = 300 + 430$, and recopied it to the right side, as can be inferred from the next extract:

I: S1, what is your proposal?
S1: $100 + 200 + 100 + 100 + 30 + 100 + 100$ [S1 verbalizes, and the Interviewer writes it on the board]
I: [...] This 100 comes from either side?
S1: Comes from 530 [S1 explains where each number comes from]
I: And for the other side, what is your expression?
S1: I took the 100 and the 200, well it is the same.

As suggested by the last line of the verbatim above, S1 just copied onto the right side her initial left-side decomposition, but she was able to indicate how the numbers of the right side related to $300 + 430$. In the same way, S3 explained his decomposition. The *ad hoc* decomposition and copying strategy that was being developed at this moment by the three students was even more evident in the work of S2. On being asked to rewrite the equality $530 + 200 = 300 + 430$, she proposed transforming only the left side to 200 + 300 + 30 + 200. The next verbatim indicates what ensued:

> I: S2, let's see what is your proposal [referring to the rewritten form for 530 + 200 = 300 + 430]
> S2: 200 + 300 + 30 [verbalizes], those numbers come from the 530.
> I: Ok, what is next?
> S2: Plus 200 [Interviewer writes on the board the expression 200 + 300 + 30 + 200]
> I: [...] And for the other side? [referring to the not-yet-decomposed right side]
> S2: 100 [verbalized]. Ah no, it should be the same, no?

As can be seen from the last line of the verbatim just above, S2's first idea to rewrite the right side would start with the number 100, but she immediately realized that it would lead to a different decomposition from its left-side counterpart, 200 + 300 + 30 + 200. After a first unsuccessful attempt at trying to relate her proposed left-side decomposition to the initial form of the right side, S2 changed her proposal: she rewrote the left side of the initial equality, 530 + 200, as 300 + 200 + 30 + 200, and for the right side simply copied the same expression – one that she could reconcile with the 300 + 430 of the right side of the initial equality.

Summarizing, the students' strategy at this moment consisted of the following: decompose the left (or right) side; explain where each number of the rewritten expression comes from with respect to the given expression; copy the rewritten expression to the right side (or left side accordingly); compare the copied expression with that of the given expression of the same side. If each number of the copied expression can be related to the initial expression of that side, then the rewritten equality is equivalent to the given equality (see the representation illustrated in Figure 3.6). We refer to this structural approach for generating equivalent numerical equalities as *"ad hoc* decomposition of one side" because only one side was being decomposed as opposed to that of genuinely decomposing both sides into a third common form.

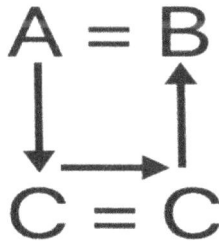

$$A = B$$

$$C = C$$

Figure 3.6 Strategy of *"ad hoc* decomposition of one side" and copying to the other side.

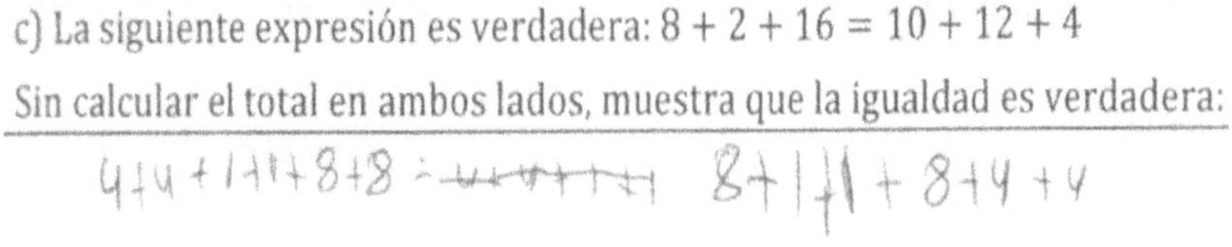

c) La siguiente expresión es verdadera: $8 + 2 + 16 = 10 + 12 + 4$

Sin calcular el total en ambos lados, muestra que la igualdad es verdadera:

4+4 + 1+1+ 8+8 = 8+ 1 /1 + 8+4 + 4

[The following expression is true: 8+2+16=10+12+4
Without calculating the total of each side, show that the equality is true]

Figure 3.7 Rewritten equality by S3.

In the application of the *ad hoc* decomposition strategy, we emphasize the important role played by students' being requested to indicate where each number of the rewritten equality came from with respect to both the left-hand and right-hand expressions of the given equality.

The Phase of Decomposition of Both Sides into a Third Common Form

As the Group Interview progressed, it became clear that the *ad hoc* decomposition strategy that included recopying the rewritten expression from one side to the other was evolving into a genuine decomposition of both sides of the initial equality so as to obtain an equivalent second equality (see Figure 3.7).

The work illustrated in Figure 3.7 suggests a structural strategy that involves the search for a *third common form*: Rewrite by decomposing both sides of the equality in a manner that shows that both sides are the same – even if the order may be different. This strategy does not involve relating both sides of the initial equality simultaneously, but rather transforming each side separately so as to obtain expressions on both sides with the same common form. As illustrated in the more generalized representation of Figure 3.8, use of the third-common-form

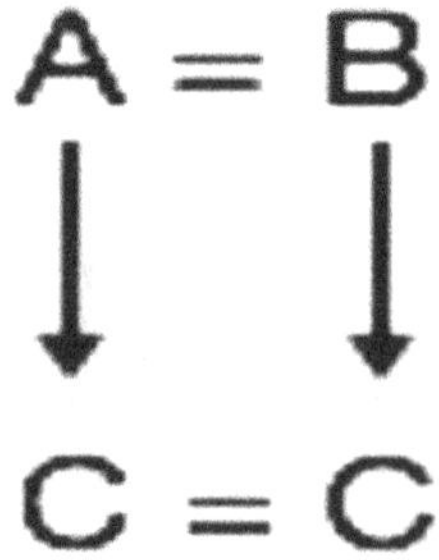

Figure 3.8 Decomposing both sides of an equality into a third common form.

property of equivalence can be expressed as follows: Given a true equality such that the left side A is written in a form that is different from that of the right side B, rewrite both sides in a common form C by means of decomposition.

This strategy based on decomposition of the numbers differs from the *ad hoc* decomposition strategy in at least two respects. First, both sides of the equality are genuinely decomposed. Second, once one side is decomposed in some form, the decomposition of the other side is guided by the first decomposition in order to get the same common form, even if the order of the terms may be different. The result is a second equality equivalent to the initial equality where the two expressions of the initial equality are now re-expressed in a common form.

The Phase of Decomposition/Composition of One Side so as to Match the Other Side of the Given Equality

An additional structural strategy was later provoked in the students' thinking when the interviewer subsequently asked them if the common form they were looking for with the third equality of the task-set (i.e., $8 + 2 + 16 = 10 + 12 + 4$) could be the left or the right side. The students then showed that they could relate both sides of the initial equality simultaneously:

> I: One more question, could this left side $[8 + 2 + 16]$ be transformed into the form of the right side $[10 + 12 + 4]$, or vice-versa?
> S1, S2, and S3: Yes [the three students at once]
> I: How would you do that?
> S1: I would do the following: the 10 can be expressed as an 8, and the 2 left [S3 interrupts]
> S3: Me, me [S3 interrupts, and wants to come forward to the board]
> I: Ok, S3 wants to show us [S3 writes what is shown in Figure 3.9]

The last step in the consolidation of the common-form strategy was thus seen with the student S3 who carried out the following transformation on the left side so as to match the right side (see Figure 3.9): Compose $8 + 2$ into 10 and decompose 16 into $12 + 4$.

Even if S3 did not rewrite the second equality including both sides, such as $10 + 12 + 4 = 10 + 12 + 4$ (maybe because of the way the

a) $8 + 2 + 16 = 10 + 12 + 4$

b) $10 + 12 + 4$

Figure 3.9 S3's composing and decomposing the left side (a) into (b) so as to match the right side.

interviewer's question was posed), his ultimate strategy involved a simultaneous relating of both sides of the initial equality.

Nevertheless, this last structural phase of decomposing one side in such a way that it matched the other side seemed a less instinctive strategy than was the decomposing-both-sides-into-a-third-common-form strategy. This became evident towards the end of the group interview session after the students had finished answering with respect to the truth-value of the three given equalities of Task-set 4 ($10 + 7 = 5 + 12$, $530 + 200 = 300 + 430$, and $8 + 2 + 16 = 10 + 12 + 4$) and they were asked to return to the equalities $10 + 7 = 5 + 12$ and $530 + 200 = 300 + 430$ and to validate them once again. The reason for this revisiting was that the students had not spontaneously used either the third-common-form or matching strategies when these two equalities were initially proposed. In giving them another opportunity to generate equivalent equalities for these two examples, we were encouraging them to show what they had learned thus far.

When the first of these two initial equalities was revisited, the following ensued:

> I: Let me go back to the first two equalities [writes on the board $10 + 7 = 5 + 12$]. How can it be re-expressed?
>
> S1 and S3: Me, me, [both want to go to the blackboard].
>
> I: S1, go ahead, please. How could you tell this is true, without adding?
>
> S1: [writes on the board $5 + 5 + 5 + 2 = 5 + 5 + 5 + 2$, see Figure 3.10, left]. This [pointing to the $5 + 5$ at the left side of the equality] I would take from the 10. This [pointing to $5 + 2$ on the left side] from the 7. This [pointing to the 5 on the right side of the equality] from 5, and this [pointing to $5 + 5 + 2$] from the 12.

As observed in Figure 3.10 (left), S1 decomposed both sides of the equality into a third common form ($5 + 5 + 5 + 2$) and explained where each number came from. In the same way, S3 proposed another decomposition, $5 + 5 + 7 = 5 + 5 + 7$ (see Figure 3.10, right), based on the same idea – also explaining where each number came from in the rewritten equality. These responses suggested that the students were spontaneously choosing the third-common-form decomposition strategy.

In the revisiting of the second equality ($530 + 200 = 300 + 430$), and as a counterproposal to the third-common-form strategy that the students

Figure 3.10 S1's decomposing strategy (left) and S3's (right).

had just used, the interviewer asked them directly about the possibility of transforming the left side into the right side (or vice versa):

> I: Finally, I am going to redo [writes $530 + 200 = 300 + 430$ on the board]. Can it be rewritten? You have already done this, but tell me how to do it in another way. This [referring to the expression $530 + 200$], for example, can it be rewritten in this way [pointing to the expression $300 + 430$]? Or this [pointing to $300 + 430$] in this way [pointing to $530 + 200$]? How would you do it?
>
> S3: Oh, yes, yes!
>
> I: Go ahead S3
>
> S1: Transforming
>
> I: Tell us S1. Transforming how?
>
> S1: By doing just as S2 and S3 just did [referring to the earlier strategy for rewriting $8 + 2 + 16 = 10 + 12 + 4$]
>
> S3: [While S1 was speaking, S3 had advanced to the blackboard] I subtract 200 from this one [pointing to 530] to make it into 300, and add the 200 to this one [pointing to 200] to make it 430. Hold on, no! 230 [S3 finally writes the equality $300 + 430 = 300 + 430$, see Figure 3.11, left].
>
> I: This [pointing to the left side $300 + 430$ written by S3], where did it come from?
>
> S3: From 530, I subtract 230 and add it to 200, so I get 430.

As indicated in the excerpt and in Figure 3.11 (left), S3's strategy was as follows: The left side is decomposed into $300 + 230 + 200$ and then the 230 and 200 are composed by addition into 430, to yield a new left-hand expression of $300 + 430$, which matches the right-hand expression of the initial equality (see also our representation of S3's approach in Figure 3.11, right). This common-form-by-matching strategy is based on a simultaneous relation that is discerned between both right and left sides of the given equality, which was not the case for the third-common-form strategy. In other words, this strategy does not involve the arbitrary decomposing of one side; rather the decomposition is directly informed by the initial form of the other side.

Figure 3.11 Rewritten equality by S3 (left); our representation of his strategy (right).

Discussion

This study has investigated how a group of 6th grade Mexican students (10–12 years of age) moved from the computational to the structural dimension of equivalence with respect to numerical expressions and equalities. Because of the strong computational orientation that they displayed in response to the tasks of Part 1, Part 2 of the study, which was conducted six months later, requested explicitly that they <u>not</u> calculate the total of each side in order to respond to the tasks. For each equality that was proposed, students were asked to rewrite the equality in such a way that showed that it was true and then to explain their reasoning. Based on the design of the tasks and the interactions that the students experienced both among themselves and with the interviewer, the development of the students' structure sense with respect to generating equivalent numerical equalities passed through five phases. The following discussion centres on the actual movement by the students from a computational to a structural way of thinking about and generating equivalent equalities, and illustrates how this movement was spurred by crucial interventions on the part of the interviewer.

Past research on algebraic thinking has provided ample evidence that structure sense does not develop spontaneously in primary school students (e.g., Linchevski & Livneh, 1999). As emphasized by Bass and Ball (2003, p. vii): "It depends on teachers who can hear the mathematics in students' talk, who can shape and offer problems of an adequate size and sufficient scope, and who can steer such problems to a productive point" – that is, it needs the thoughtful intervention of a knowledgeable teacher. This study was no exception. Even well-designed tasks in and of themselves do not manage the educational business of fostering the growth of algebraic thought (Sullivan et al., 2015; see also Sáenz-Ludlow & Walgamuth, 1998). The excerpts that were presented in the above section have illustrated the ways in which the interviewer's prompts and manner of responding to the students' efforts were crucial to leading them from the computational to the structural dimension of equivalence for numerical expressions and equalities. To highlight the interventions that were key to this development, we have synthesized within Table 3.1 not only the principal interventions but also the way in which these precipitated the evolution in students' thinking.

As is suggested by the summary data of Table 3.1, the third phase of *Ad Hoc* Decomposition of One Side and Copying to the Other Side, where five crucial interviewer interventions occurred, was the lengthiest of the five phases and the one that was pivotal to the development of students' structure sense.

- The first intervention of the third phase, not surprisingly, was the request to show that the given equalities were either true or false, but without calculating the total for each side.

Table 3.1 Summary of Key Intervention Prompts and Resulting Student Actions in the Evolution of Their Structure Sense for Equivalence of Expressions and Equalities

Phases in the Development of Students' Structure Sense	Intervention Prompts by the Interviewer	Students' Reactions
1. Computation Without Decomposition (in Part 1 of the Study)	*Is the given equality statement true or not and explain why?*	Students justified the truth (or falsity) of each of the five given equality statements by computing the result on each side – whether the equality involved small or large numbers.
2. Computation with Unrelated Decompositions of Both Sides (in Part 1 of the Study)	*In what other way could you rewrite the given equality so as to show that it is true?*	Students decomposed each expression of the given equality in a manner that conformed to their already computed total for each side, but not in a way that showed the relationship of sameness between the two sides of the equality. They looked for two (or more) different numbers for one side of the equality that would yield the already-computed total and another two (or more) different numbers for the other side.
3. *Ad Hoc* Decomposition of One Side and Copying to the Other Side (in Part 2 of the Study)	i. <u>*Without calculating the total of each side*</u>, show that the equality is true. ...	Looks of puzzlement
	ii. *Would there be a way to rewrite but using, say, these same numbers [points to the board where the equality $10+7=5+12$ is written]* ...	A student decomposed left side into $5+5+5+2$ and subsequently copied that expression onto the right side. (A follow-up equality involving larger numbers was similarly handled.)
	iii. *Does this 5 [points to the first number 5 on the left side, $5+5+5+2$, of the equality] come from any part here [points to the initial equality $10+7=5+12$]?* ...	When asked where the numbers of the first decomposed expression of the left side of the equality came from, the student was able to explain how he had carried out the decomposition of that expression so as to generate its transformed version.
	iv. *How would you justify these [referring to the right side of the equality after the left side had been decomposed and copied to the right side]?* ...	Students related the decomposed terms of the right side to the corresponding expression of the initial equality. (If they were unable to relate the decomposed expression to each side of the initial equality, they changed their decomposition.)
	v. *Seeing it this way [with the same expression on each side of the resulting equality], is it necessary to calculate the total [of each side]?*	Students said that, "*if the expression is the same [on both sides], if will be equal, it will give the same.*"

(Continued)

Table 3.1 Summary of Key Intervention Prompts and Resulting Student Actions in the Evolution of Their Structure Sense for Equivalence of Expressions and Equalities *(Continued)*

Phases in the Development of Students' Structure Sense	Intervention Prompts by the Interviewer	Students' Reactions
4. Decomposition of Both Sides into a Third Common Form (in Part 2 of the Study)	No special interventional prompt; this was a spontaneous evolution that followed from the prompts and reactions of the previous phase.	Students began to transform each side separately so as to obtain expressions on both sides with the same common form, even if the order of the addends might be different.
5. Decomposition/ Composition of One Side so as to Match the Other Side of the Given Equality (in Part 2 of the Study)	*Could this left side [8+2+16 of the equality 8+2+16=10+12+4] be transformed into the form of the right side [10+12+4], or vice-versa?*	Students simultaneously related both sides of the initial equality in such a way that there was no arbitrary decomposing of one side; rather the decomposition was informed by the form of the other side of the initial equality.

- This needed to be supplemented by the suggestion, implicit in the interviewer's follow-up question, that asked if the students could find a way to rewrite the given equality, but based on the numbers that were in the equality. This suggestion was central to the students' beginning to move in a structural direction.
- When the request to rewrite the given equality in a way that was based on the numbers in the equality led to students' decomposing, first, one of the two expressions of the given equality, they were asked explicitly where the decomposed numbers came from, that is, how the decomposed numbers related to the terms of the original expression on one side of the given equality. This question was crucial to developing students' awareness of the structural relationship between the expression on one side of a given equality and its decomposed version on the same side of the transformed equality.
- Since the students then simply recopied the decomposed expression over to the other side of the equals sign in order to complete the transformed equality, the interviewer asked them to justify their recopied expression, that is, to relate explicitly the recopied expression to the corresponding expression of the initial equality. This prompt was especially important in that it led students who were not able to relate the copied, decomposed expression to the corresponding expression of the initial equality to formulate a second *ad hoc*

decomposition of one of the sides – one that they would in turn be able to justify with respect not only to the corresponding expression of the initial equality but also to its partner expression on the other side of the equals sign.

- And the last prompt of this third phase – in the face of seeing a resulting equality with both sides consisting of the same expression – involved eliciting from the students the explicit and stated awareness that they no longer needed to total each side of an equality in order to determine and show its truth-value.

From this third phase onward, the realization on the part of students that equivalence could be shown by converting either/both expressions of an equality into a common form was actualized.

While the point of departure for the group interactions was the tasks themselves, it should be added that the students had worked individually on these and had thought about how to answer them before being asked to share their approaches and their thinking. But the tasks cannot be separated from the way in which the interviewer posed additional queries about them and verbally extended the actual written questions that had been integral to the tasks. Because the interviewer had such a deep understanding of the tasks and their rationale, and had in fact co-constructed the tasks along with the first author, he knew how to steer the student discussion of the tasks to a productive point. Thus, the combination of tasks, interviewer interventions, and students' sharing of their thinking all worked together to effectively move the students from the computational to the structural dimension of equivalence. In sum, the development of structure sense at this level of schooling depended on these cultural interactions and surely would not have happened without them.

In the movement of the students to the structural dimension of equivalence, three features of a cognitive nature are quite noteworthy. They concern: (i) the role played by the decomposing property of the addition structure in the students' evolution, (ii) the way in which their use of properties, especially that of decomposition, was grounded in their sense of number and computational knowledge, and (iii) their inclination to the third-common-form approach for generating equivalent equalities in contrast to the common-form-by-matching-one-of-the-sides approach.

The interaction between the first two of these features was particularly remarkable in view of the fact that the students were not to compute the total of each side in order to show the truth-value of the equalities. First, we noted that the way in which the decompositions of the various numerical terms of the expressions were written showed certain numerical structures – a manner of re-expressing that emerged directly from the students. For example, in Phase 2, S1 decomposed $172 + 10 + 75$

into 100 + 72 + 5 + 5 + 60 + 15, and 182 + 50 + 25 into 100 + 82 + 30 + 20 + 20 + 5. In Phase 3, S3 decomposed 10 + 7 into 5 + 5 + 5 + 2, and in Phase 4, 8 + 2 + 16 into 4 + 4 + 1 + 1 + 8 + 8; while S2 in Phase 3 decomposed 530 + 200 into 300 + 200 + 30 + 200. The presence of basic units of place-value such as 100s and 10s, as well as for S3 the use of doubles and halves, is quite noticeable. This led us to then conjecture that, during their years of primary schooling, these students had been able to develop a rather strong number sense in the growing of their computational knowledge. Not only could they, as expected, compute forward (as was seen in their initial approaches to determining the truth-value of equalities), but they could also work backwards by breaking up numbers into their principal structural parts – even if this was not their first reflex. While the transitive property of equality may also have been implicit in the students' thinking, as well as the reflexive property of equality (i.e., $a = a$, where a is a numerical expression), it was the property of decomposition of the addition structure – well bolstered by the students' sense of numerical relationships – that was salient throughout Phases 2 to 5 in the development of their structure sense.

The role played by knowledge of numerical relations in the students' building of structure sense for equivalent equalities touches indirectly on a point made by Baroody and Ginsburg (1983). Based on their research with younger students, they argued that they would not try to eliminate children's initial tendencies to compute in order to establish equality and to make sense of the equals sign – and that this knowledge is vital to making sense of later instruction designed to advance their computationally-oriented approaches towards more structural conceptions. The fact that knowledge of numerical structure and aspects of computation provide the underpinnings for the development of students' structural sense for equivalent equalities also resonates with Sfard's (1991) theory of operational-structural duality. As Sfard (1994, p. 53) argues: "From the developmental point of view, operational conceptions precede structural." The importance of a strong numerical base for the use of the decomposing property leads to a related suggestion regarding students' later work in school algebra: Their well-documented difficulties with correctly applying properties in the structurally-based activity of equation solving might benefit greatly from having more time spent with the numerical footings of these properties in the early stages of learning algebra. While we do not, in any way, minimize the importance of the structural dimension of equivalence that we fostered with our study, we would be remiss if we did not raise the point about the role played by students' number sense and computational knowledge in making their transition to the structural.

Last, a few remarks are in order regarding the seeming inclination of the students for the third-common-form rather than the common-form-by-matching-one-of-the-sides approach for generating equivalent equalities, which was revealed during the fifth phase. The common-form property of equivalence states that, for a pair of expressions, converting either/both expressions into a common form can be used to indicate equivalence. Recall that, in the latter segment of the fifth phase, the already-worked-on equalities were revisited; this was just after the students had experienced both approaches to generating equivalent equalities. When the first equality was revisited, the students spontaneously decomposed each side into a third common form. For the next equality, the interviewer asked them if they might instead use the matching-one-of-the-sides approach. We hypothesize that the simultaneous relating of both sides of a numerical equality so as to use the matching approach may be more cognitively demanding for students of this age than the approach of decomposing one side in some form and then using this form to guide the decomposition of the other side in order to obtain a third form that is common. While the parallel processing required for the matching approach likely requires more cognitive resources than an approach involving term-by-term decomposing, clearly further research would be required in order to cast more light on this particular facet of students' generating of equivalent numerical equalities.

Concluding Remarks

This chapter has offered a review of the theoretical and empirical literature related to equivalence and structure, with attention to the properties associated with the structural dimension of equivalence of numerical equalities. It has also provided a detailed description of the phases that the 10- to 12-year-old students who participated in the study passed through in evolving their thinking from the computational to the structural dimension of equivalence. The discussion of the results focused in particular on the fundamental role played by the interviewer's interventions in the development of students' structure sense for generating equivalent numerical equalities, as well as on the nature of the decomposing property that characterized students' structuring activity and the accompanying support offered by their sense of numerical structure and relationships. To conclude, changing an equality's form by means of decomposition was found to be central to generating equivalent numerical equalities, just as it is central to generating equivalent algebraic expressions and equations. With the portrait that we have presented of the manner in which students' thinking can be guided to transition towards structural considerations of numerical equivalence, we hope to contribute not only to research and practice on ways to encourage the growth of primary-level students' structure sense, but also to fostering greater awareness of the crucial underpinnings of secondary-level students' algebra structure sense.

Note

1 Cesar Martínez-Hernández's life was sadly taken away from him by COVID-19 on December 15, 2020. I am immensely grateful for the altogether too brief period during which we were able to collaborate. His contributions were crucial to the unfolding of the research presented in this chapter.

References

Arcavi, A., Drijvers, P., & Stacey, K. (2017). *The learning and teaching of algebra: Ideas, insights, and activities.* London: Routledge.

Asghari, A. (2019). Equivalence: An attempt at a history of the idea. *Synthese, 196*(11), 4657–4677. https://doi.org/10.1007/s11229-018-1674-2

Asghari, A. H., & Khosroshahi, L. G. (2017). Making associativity operational. *International Journal of Science and Mathematics Education, 15*(8), 1559–1577.

Baroody, A. J., & Ginsburg, H. P. (1983). The effects of instruction on children's understanding of the "equals" sign. *Elementary School Journal, 84*(2), 198–212.

Bass, H., & Ball, D. L. (2003). Foreword. In T. P. Carpenter, M. L. Franke, & L. Levi (Authors), *Thinking mathematically: Integrating arithmetic and algebra in elementary school.* Portsmouth, NH: Heinemann.

Blanton, M. L., Brizuela, B. M., Stephens, A., Knuth, E., Isler, I., Gardiner, A. M., Stroud, R., Fonger, N. L., & Stylianou, D. (2018). Implementing a framework for early algebra. In C. Kieran (Ed.), *Teaching and learning algebraic thinking with 5- to 12-year-olds: The global evolution of an emerging field of research and practice* (pp. 27–49). New York: Springer.

Carpenter, T. P., Franke, M. L., & Levi, L. (2003). *Thinking mathematically: Integrating arithmetic and algebra in elementary school.* Portsmouth, NH: Heinemann.

Freudenthal, H. (1983). *Didactical phenomenology of mathematical structures.* Dordrecht, NL: Reidel.

Freudenthal, H. (1991). *Revisiting mathematics education: China Lectures.* Dordrecht, NL: Kluwer Academic.

Hoch, M., & Dreyfus, T. (2004). Structure sense in high school algebra: The effect of brackets. In M. J. Høines, & A. B. Fuglestad (Eds.), *Proceedings of 28th Conference of the International Group for the Psychology of Mathematics Education* (Vol. 3, pp. 49–56). Bergen, NO: PME.

Jacobs, V. R., Franke, M. L., Carpenter, T. P., Levi, L., & Battey, D. (2007). Professional development focused on children's algebraic reasoning in elementary school. *Journal for Research in Mathematics Education, 38*(3), 258–288.

Kaput, J. J. (2008). What is algebra? What is algebraic reasoning? In J. J. Kaput, D. W. Carraher, & M. L. Blanton (Eds.), *Algebra in the early grades* (pp. 5–17). New York: Lawrence Erlbaum.

Kieran, C. (1981). Concepts associated with the equality symbol. *Educational Studies in Mathematics, 12*(3), 317–326.

Kieran, C. (1989). The early learning of algebra: A structural perspective. In S. Wagner & C. Kieran (Eds.), *Research issues in the learning and teaching of algebra* (pp. 33-56). Reston, VA: National Council of Teachers of Mathematics.

Kieran, C. (2007). Learning and teaching algebra at the middle school through college levels: Building meaning for symbols and their manipulation. In F. K. Lester (Ed.), *Second handbook of research on mathematics teaching and learning* (pp. 707–762). Greenwich, CT: Information Age.

Kieran, C. (2018). Seeking, using, and expressing structure in numbers and numerical operations: A fundamental path to developing early algebraic thinking. In C. Kieran (Ed.), *Teaching and learning algebraic thinking with 5- to 12-year-olds: The global evolution of an emerging field of research and practice* (pp. 79–105). New York: Springer.

Kieran, C., & Martínez-Hernández, C. (2021). Developing structural thinking for equivalence of numerical expressions and equalities with 10- to 12-year-olds. In M. Inprasitha, N. Changsri, & N. Boonsena (Eds.), *Proceedings of the 44th Conference of the International Group for the Psychology of Mathematics Education* (Vol. 3, pp. 128–137). Khon Kaen, Thailand: PME.

Knuth, E. J., Stephens, A. C., McNeil, N. M., & Alibali, M. W. (2006). Does understanding the equal sign matter? Evidence from solving equations. *Journal for Research in Mathematics Education*, 37(4), 297–312.

Lee, J., & Pang, J. (2021). Students' opposing conceptions of equations with two equal signs. *Mathematical Thinking and Learning*, 23(3), 209–224.

Li, X., Ding, M., Capraro, M. M., & Capraro, R. M. (2008). Sources of differences in children's understandings of mathematical equality: Comparative analysis of teacher guides and student texts in China and the United States. *Cognition and Instruction*, 26(2), 195–217.

Linchevski, L., & Livneh, D. (1999). Structure sense: The relationship between algebraic and numerical contexts. *Educational Studies in Mathematics*, 40(2), 173–196.

Martínez-Hernández, C., & Kieran, C. (2018). Strategies used by Mexican students in seeking structure on equivalence tasks. In T. E. Hodges, G. J. Roy, & A. M. Tyminski (Eds.), *Proceedings of the 40th Annual Meeting of PME-NA* (pp. 163–170). Greenville, SC: University of South Carolina & Clemson University.

Martínez-Hernández, C., & Kieran, C. (2019). From computational strategies to a kind of relational thinking based on structure sense. In S. Otten, A. G. Candela, Z. Araujo, C. Haines, & C. Munter (Eds.), *Proceedings of the 41st Annual Meeting of the North American Chapter of the International Group for the Psychology of Mathematics Education* (pp. 167–176). St. Louis, MO: University of Missouri.

Martínez-Hernández, C., & Kieran, C. (2020a). Decomposing, composing, and recomposing numbers in numerical equalities: Algebraic thinking based on structure sense / Descomposición, composición y recomposición de números en igualdades numéricas: Pensamiento algebraico basado en un sentido de estructura. In A. I. Sacristan, J. C. Cortés-Zavala, & P. M. Ruiz-Arias (Eds.), *Proceedings of the 42nd Annual Meeting of the North American Chapter of the International Group for the Psychology of Mathematics Education* (pp. 305–313). Mazatlan, Mexico: PME-NA.

Martínez-Hernández, C., & Kieran, C. (2020b). The development of algebraic thinking in primary school: From computational to structural strategies. *Paper accepted for presentation at the 14th International Congress on Mathematical Education*, Shanghai, China.

Mason, J., Stephens, M., & Watson, A. (2009). Appreciating mathematical structure for all. *Mathematics Education Research Journal, 21*(2), 10–32.

Mason, J., with Burton, L., & Stacey, K. (2010). *Thinking mathematically* (2nd ed.). London: Pearson.

Molina, M., & Ambrose, R. C. (2006). Fostering relational thinking while negotiating the meaning of the equals sign. *Teaching Children Mathematics, 13*(2), 111–117.

Radford, L. (2018). The emergence of symbolic algebraic thinking in primary school. In C. Kieran (Ed.), *Teaching and learning algebraic thinking with 5- to 12-year-olds: The global evolution of an emerging field of research and practice* (pp. 3–25). New York: Springer.

Rittle-Johnson, B., Matthews, P. G., Taylor, R. S., & McEldoon, K. L. (2010). Assessing knowledge of mathematical equivalence: A construct modeling approach. *Journal of Educational Psychology, 103*(1), 85–104.

Sáenz-Ludlow, A., & Walgamuth, C. (1998). Third graders' interpretations of equality and the equal symbol. *Educational Studies in Mathematics, 35*(2), 153–187.

Schifter, D. (2018). Early algebra as analysis of structure: A focus on operations. In C. Kieran (Ed.), *Teaching and learning algebraic thinking with 5- to 12-year-olds: The global evolution of an emerging field of research and practice* (pp. 309–328). New York: Springer.

Secretaría de Educación Pública (S. E. P.). (2016). *Propuesta curricular para la educación obligatoria 2016.* México, DF: SEP.

Sfard, A. (1991). On the dual nature of mathematical conceptions: Reflections on processes and objects as different sides of the same coin. *Educational Studies in Mathematics, 22*(1), 1–36.

Sfard, A. (1994). Reification as the birth of a metaphor. *For the Learning of Mathematics, 14*(1), 44–55.

Sullivan, P., Knott, L., & Yang, Y. (2015). The relationships between task design, anticipated pedagogies, and student learning. In A. Watson, & M. Ohtani (Eds.), *Task design in mathematics education – An ICMI study 22* (pp. 83–114). New York: Springer.

Venkat, H., Askew, M., Watson, A., & Mason, J. (2019). Architecture of mathematical structure. *For the Learning of Mathematics, 39*(1), 13–17.

Venkat, H., Beckmann, S., with Larsson, K., Xin, Y. P., Ramploud, A., & Chen, L. (2018). Connecting whole number arithmetic foundations to other parts of mathematics: Structure and structuring activity. In M. G. Bartolini Bussi, & X. H. Sun (Eds.), *Building the foundation: Whole numbers in the primary grades* (pp. 299–324). New York: Springer.

Warren, E. (2003). The role of arithmetic structure in the transition from arithmetic to algebra. *Mathematics Education Research Journal, 15*(2), 122–137.

4 Developing Structure Sense with Digital Technologies

Introducing the MEx Platform

Valentina Muñoz-Porras
and Ulises Xolocotzin

Introduction

Ever since computers became widely available for educational purposes, researchers and teachers have been interested in finding ways of creating and using digital tools for supporting the teaching and learning of algebra. There are several tools and environments available to investigate and support different aspects of algebraic thinking. However, more research is necessary to find ways of using digital technology to support the development of symbolic algebra structure sense – we will be calling this structure sense throughout this chapter. An operational view of algebra is insufficient for accessing the level of mathematical thinking required to succeed in higher-education mathematics, for example, in STEM careers. Therefore, students need to acquire structure sense in high school.

In this volume, Rojano elaborates on what structure sense means in the context of high-school mathematics. In summary, structure sense is a set of skills that allows students to (i) recognise a familiar structure in its simplest form, (ii) deal with a compound term as a single entity, (iii) through an appropriate substitution, recognise a familiar structure in a more complex form, and (iv) choose appropriate manipulations to make the best use of a structure (Hoch & Dreyfus, 2005). These are all critical aspects of symbolic algebra that can contribute to developing the advanced mathematical competencies required in higher education. Therefore, algebra education scholars emphasise the need to create learning opportunities for high school students to develop structure sense.

Structure sense is a highly sophisticated skill that involves abstract ways of thinking about symbolic algebra. Students cannot learn structure sense in the same ways they learn other aspects of algebra. Typical curricula emphasise three conceptions of algebra: generalised arithmetic, problem-solving procedures, and relationships between quantities. Therefore, most digital tools designed to support algebra learning address these notions, which is, of course, necessary. However, we agree with Kieran in that *"equal attention needs to be paid to the complementary*

DOI: 10.4324/9781003197867-4

*process of looking through mathematical objects [...] and to decompos-
ing and recomposing them in various structural ways*" (Kieran, 2018,
p. 80). Unfortunately, there is very little research concerning the design of
digital technologies for developing structure sense in high-school algebra.
This chapter presents some design principles to support algebra structure
sense. We review principles like meaningful practice, gamification, and
adaptivity. We will also introduce the MEx platform. Instead of covering
the typical high-school mathematics curricula, MEx promotes structure
sense through powerful algebraic ideas, like equivalence, substitution, and
the object-process duality of algebraic expressions.

Teaching and Learning Algebra with Digital Technologies

There are different approaches to the use of digital technologies in the
teaching and learning of algebra. A thorough review of these approaches
and the technologies that represent them is beyond the scope of this
chapter. Instead, we aim to discuss the design principles and learning
potential of some representative technologies designed with the specific
purpose of easing the acquisition of algebraic skills, with an emphasis on
the development of structure sense. Admittedly, we miss the illuminat-
ing teaching and learning experiences that emerge from using domain-
general technologies to teach algebra in the high-school mathematics classroom. Examples of these technologies include calculators and
interactive whiteboards, learning platforms such as Google classroom
and Kahoot, or popular collaboration tools like Peardeck and Google
Jamboard. However, we considered that focusing on technologies designed
with an algebraic perspective could be helpful to identify features that
can support the development of structure sense. Technologies can ease
algebraic activities – for example, accelerating computations, illustrating symbolic manipulation procedures, or facilitating decisions in problem-solving. However, our discussion is not about these aspects of digital
technology. Instead of this, we try to elucidate how digital technologies
might support the <u>acquisition</u> of algebraic skills like structure sense.

Designing with Users in Mind: Tools for Teachers and Tools for Learners

Digital technologies designed with an algebraic perspective offer tools
that support the complementary processes of teaching and learning.
Some technologies support teachers in the design of didactical activities. We call these teacher-oriented tools. Other technologies offer tools
designed to support learners' cognitive activity without necessarily rely-
ing on the mediation of teachers. We call them learner-oriented tools. We
recognise that learner-oriented and teacher-oriented technologies often

share functionalities. However, we believe that identifying the design orientation of a given tool is helpful to see how its features can support specific aspects of students' algebraic thinking.

There is a range of environments and platforms with teacher-oriented tools. These tools tend to be open-ended and flexible because they support teachers in designing tasks and the organisation of didactical sequences. For example, in the Polypad App of the Mathigon platform (https://mathigon.org), teachers can use visual non-symbolic representations like algebra tiles or balance scales for designing activities involving operations with integers and variables. Another widely known example is Geogebra (https://www.geogebra.org), which includes an interactive geometry system, spreadsheets, and a computer algebra system (CAS). These features offer endless possibilities in the creation of applets to teach about specific aspects of algebra. The typical use of Geogebra consists of making students interact with an applet and, with the guidance of teachers, making conjectures and argumentations.

Unlike teacher-oriented tools, learner-oriented tools support the development of algebraic skills directly. Often, learner-oriented tools are part of larger learning environments or platforms. Examples include the tutorial videos and practice exercises embedded in Khan Academy (https://khanacademy.org) or the solved examples of Alcumus (https://artofproblemsolving.com/alcumus). Learner-oriented tools can also be part of stand-alone apps or games. Such is the case of the different metaphors and mechanics designed for learners to work with syntactic rules without direct instruction, such as the mobile game Dragon Box (https://dragonbox.com). Another example is the use of function machines that integrate algebraic substitutions in the MEx platform that we will present below.

Designing Algebraic Experiences: Making Interactions Based on Algebraic Principles

The ever-accelerating progress of digital technologies constantly opens innovative possibilities for re-thinking the design of educational tools. Here we discuss some developments in interaction design for teaching and learning algebra. We focus on interface design, user-generated content, game-based learning, multi-representational learning, and symbolic algebra support.

Interface Design

The advent of new interfaces allows new ways of supporting the learning of algebra. The way that we interact with computers has changed dramatically over the last two decades. The first digital tools employed in the algebra classroom were part of devices like CAS calculators and the early models of desktop computers. Albeit innovative, these technologies were inaccessible for most students and teachers. Nowadays, powerful and affordable devices like tablets and smartphones are available for mass consumption. This evolution in hardware brought graphics and touch interfaces to algebra learning software. These interfaces allow more intuitive interactions that guide the actions and reasoning of learners with affordances, movement constraints, and visual guides. Graspable Math (https://graspablemath.com/) is an excellent example of the current possibilities of combining graphic and touch interfaces for supporting symbolic algebra activities. In Graspable Math, algebraic expressions are dynamic. The learner can manipulate them according to the syntactic rules of algebra (Weitnauer et al., 2017). We argue that this is a game-changing approach to interaction design. Pioneering technologies like CAS calculators allowed the typing of algebraic expressions with a keypad. However, these expressions remained static because the CAS interface did not permit the direct manipulation of symbols.

User-Generated Content

The emergence and eventual dominance of technologies that allow users to generate and share content changed radically educational technologies. The quantity and quality of user-generated content have expanded rapidly. The increasing number of YouTube content for learning and teaching algebra is probably the best example of this. The nowadays massively used Khan Academy (https://www.khanacademy.org/) started in this way. Around 2006, the founder, Sal Khan, began uploading mathematics tutorials for his family on YouTube. By 2021, Khan Academy reached about 50 million users across 190 countries. A sceptic reader might argue that user-generated tutorial videos are nothing more than amplified versions of traditional teaching. Tutorial videos are, after all, a form of direct instruction in which an instructor mediates the interaction between learners and mathematical objects. However, digital technologies do not magnify learning. Instead, they reorganise how it is exercised (Crook, 2001), and tutorial videos reorganise how instruction works. The learner can manipulate a lesson, pausing it or rewinding it for revisiting or reflecting about content to achieve understanding. It is almost needless to say that this is impossible without technology. Moreover, in platforms like Khan Academy or Alcumus (https://artofproblemsolving.com/alcumus), tutorial videos combine with practice exercises that support active learning.

Game-Based Learning

The idea of using computer games to support the learning of mathematics has been around since computers entered the mathematics classroom (Xolocotzin & Pretelín-Ricárdez, 2015). The influence of the gaming culture in the conceptualisation of learning technologies is constantly increasing, which opens new possibilities for designing algebra learning tools. Most educational games follow a "chocolate-covered broccoli" approach (Habgood & Ainsworth, 2011). That is, adding a layer of game features on top of rote learning exercises. However, when properly integrated with mathematical foundations, the principles of game design can lead to powerful learning experiences. Dragon Box (https://dragonbox. com/products/algebra-5) is an example of the successful integration of game mechanics with algebraic foundations. This game helps children to learn how to solve first-degree equations. The goal of the game is to equate boxes. There are boxes representing integers and boxes representing unknown quantities. The game introduces symbols like literals progressively. The rules for equating the boxes follow the algebraic rules for solving equations. Gaming elements including immediate feedback, rewards, and increasing difficulty help sustain the learner's attention in meaningful practice tasks. Other examples of games that integrate game mechanics with mathematics include Euclidea (https://www.euclidea. xyz) and Sinerider (https://sinerider.com).

Multi-Representational Learning

Learners can represent algebraic structures in various ways, for example, with notation, graphs, and pictures. There is a long tradition of algebra learning tools that integrate multi-representational features. Some technologies allow interactions with formal algebraic representations, for example, Desmos (https://www.desmos.com/calculator) and Geogebra (https://www.geogebra.org) integrate graphs, tables, and symbolic expressions. In Desmos, the user can create and classify cards representing the same algebraic object in different ways. Desmos also has an excellent editor for adding mathematical text to activities that can also include graphs or tables. Other technologies allow interactions with non-formal visual representations of algebraic objects. For example, in the Polypad App of the Mathigon platform (https://mathigon.org), the user (the teacher) can build activities with manipulative tiles that represent algebraic structures. Some applets of the platforms Phet (https:// phet.colorado.edu/en/simulation/equality-explorer) and Mathigon's Polypad use metaphorical representations like scale balances for supporting the notion of equality in algebra.

Symbolic Algebra Support

One critical development in algebra learning technologies is the integration of tools for working with algebraic text symbols. Most technologies have a keyboard for this purpose. Moreover, recent apps commonly used in the mathematics classroom, like Photomath (https://photomath.com)

and MS Math Solver (https://math.microsoft.com), transform the photograph of an expression into computerised text and offer tools for working with it, for example, step-by-step explanations of the algebraic procedures required for solving it. MS Math Solver also lets the user draw the mathematical expression with a stylus and transform it into computerised math text. Jupri and Sispiyati (2020) analysed procedural fluency and conceptual understanding in algebra. They let students use Photomath to solve symbolic algebra tasks. Step-by-step solutions could help students improve their algebraic skills but privileged procedural strategies over structure sense.

Supporting Structure Sense with Algebra Learning Tools

This chapter introduces the MEx platform. To the extent of our knowledge, this is the first technology designed with the specific purpose of supporting the development of symbolic algebra structure sense. MEx takes advantage of tools that are present in currently available algebra learning tools. *Structure sense* is an umbrella term for skills that allow identifying and processing algebraic structures represented in different symbolic expressions. Below we present some examples of how the tools integrated into some of the algebra learning technologies mentioned above can support the development of structure sense. We made these examples from a teaching perspective. Since practice is essential for developing structure sense, teachers can design and orchestrate a series of sessions with teacher-oriented tools like Desmos or Mathigon's polypad. Another route is to take advantage of learner-oriented tools, like the videos and activities available on Khan Academy and Alcumus. Finally, we would like to propose using the Mex platform, described later in this chapter.

Playing with Desmos Marbleslides

Marbleslides (https://learn.desmos.com/marbleslides) is a game developed in Desmos, a platform with learning and teaching tools based on the principles of a graphic calculator. The mechanics of Marbleslides are straightforward. The user sees a graph with a curve representing an algebraic expression written in a panel in the app window. Above the curve, there are some stars. When pressing the "Launch" button, some marbles will fall from the upper right. The goal of the game is to make the marbles get as many stars as possible. The marbles fall attracted by gravity and will slide above the curve. The player can change the shape of the curve by changing the associated expression. Therefore, the game is about predicting and controlling the behaviour of the marbles so they can get the stars, a rewarding and enjoyable activity. The system gives immediate feedback. The game promotes intuitions about the relationship between the shape of a curve and the expression it represents.

Marbleslides offers sequences of ready-to-use tasks, but teachers can also modify the equations and position of the stars to create their sequences of tasks. The ready-to-use Marbleslides are lines, parabolas, rationals, periodicals, and exponentials. There is scaffolding in each sequence because solving the problem is presented in the correct form. For example, in Marbleslides Lines, the curves are lines of the form $y = mx + b$. An instance of that equation that draws a line, say $y = 2x + 3$, the user needs to change something in the equation to solve the puzzle. There is also levelling-up because, first, puzzles are solved by changing something simple in the expression, for example, replacing number 3 with number 2. However, some other puzzles need more changes in parts of the expression. This game dynamic promotes focusing on sub-expressions or parts of the equations and relating them to changes in the associated curve. The objective is not memorising the effects of changing m and b in the graph of a linear equation. Instead, the objective is to help students educate their intuitions as they progress in the sequence of puzzles. We argue that this kind of thinking could promote structure sense, although further studies are needed.

Dynamic Symbol Manipulation with Graspable Math

Graspable Math is one of the few technologies focused on the symbolic manipulation of algebraic expressions – see Aplusix (https://aplusix. org/) for another example. Graspable Math allows the manipulation of algebraic notation with manipulative gestures such as drag and drop. The manipulation of symbolic elements, for example, variables, parentheses, exponents, is a distinctively algebraic functionality of Graspable Math. Graspable Math constrains the manipulation of symbols with restrictions aligned with the syntactic procedures of symbolic algebra. Suppose the user moves a symbol or a subexpression in an invalid way. In that case, the change does not occur, and the symbol or the sub-expression goes back to its original position. In this way, Graspable Math integrates algebraic structures with the mechanics of its interface. We believe that this is a remarkable achievement that illustrates what designing with a structural perspective means. Graspable Math also offers tools that teachers can employ. For example, the history of the manipulations made by a student while solving a task is available, and the teacher can consider it for understanding the students' way of thinking. Teachers can also make a sequence of tasks inside the platform to promote specific ways of thinking. For example, one can ask the student to rewrite part of an expression, so the formula module recognised it as a known structure.

Graspable Math implements a range of tools that can be employed to support the development of structure sense. Some of these tools are gestures like the "shaking" of expressions or subexpressions. The shaking gesture changes an expression into anything equivalent. Another tool is the formula button. When clicked, a list with formulas is displayed. Dragging and dropping a formula on an expression in the canvas produces a visual highlighting of the text. Suppose the dragged expression has a detectable structure. If so, Graspable Math colours it in blue. Of course, experts can detect structures that Graspable Math cannot detect. For example, Graspable Math can identify a perfect square trinomial in the expression $(x+1)^2 + 2(6)(x+1) + 6^2$, but not in the expression $(x+1)^2 + 12(x+1) + 6^2$. Another example; it can detect $x^2 - 5^2$ as a difference of squares but not $x^2 - 25$. Graspable Math also affords the user to make a substitution of one variable by an expression. For example, one can write on the canvas an equation like $a = x + 2$ and the expression $a^2 + 2a + 1$. If a student drags the literal a from the equation onto the expression, Graspable Math will transform the latter into $(x+2)^2 + 2(x+2) + 1$. This functionality supports substitution skills that are essential for promoting structure sense in students. We will discuss the importance of substitution for structure sense below when describing the algebraic foundations of MEx.

Manipulating Algebraic Visual Representations in Mathigon

Working with multiple representations of the same algebraic structure eases algebraic thinking processes. Therefore, the use of multiple representations is a desirable feature of algebra learning technologies. Teachers in the algebra classroom widely employ algebra tiles. Mathigon's Polymath allows the design of activities to drag a graphic object to a canvas and arrange them by rotation or flipping. There is a dashboard where teachers can view learners progressing in the activity. Teachers use algebra tiles for teaching about factorisation and distributivity. However, Caglayan (2014) notes that teachers can also use algebra tiles to promote relational thinking. For example, they can encourage students to make "sum = product" connections, interpreting the area of a rectangle both as a sum and as a product. Learners can understand structures by associating different yet equivalent expressions to the same arrangement of tiles forming a rectangle. For example, in Figure 4.1, students can recognise several equivalent expressions representing the area of that rectangle and even associate a natural language description to that expression. Teachers can go further and create sorting games and ask students to classify equivalent expressions, for example, sorting cards with structures expressed in natural language, symbols, or areas. Rojano (Chapter 1) and Kieran (Chapter 3) noted that equivalent

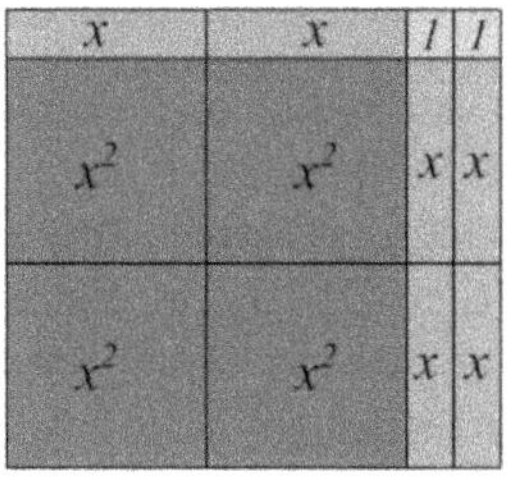

Expression	Natural language description
$(2x + 2)(2x + 1)$	Length times width
$2(2x^2 + 2x) + (2x + 2)$	Two rows of area $2x^2 + 2x$ plus one row of area $2x + 2$
$2(2x^2 + x) + 2(2x + 1)$	Two columns of area $2x^2 + x$ plus another two columns of area $2x + 1$

Figure 4.1 Equivalent expressions with algebra tiles. Different ways of seeing the same area have different yet equivalent associated expressions.

expressions play an essential role in promoting structure sense, as discussed in the next section.

Solving Structure Problems in Khan Academy and Alcumus

Tools for supporting the development of structure sense are rare. However, platforms designed to deliver algebra curricula do include structure sense components. For example, Khan Academy supports meaningful practice with problems tailored to specific mathematics domains and tutorials. In Khan Academy, there are about a thousand exercises designed to cover the "Seeing structure sense" theme of the US Common Core Standards. Alcumus (https://artofproblemsolving.com/alcumus) is a platform that can also support structure. Alcumus is an open-access platform with a collection of about 13,000 Mathematics Olympiad problems, including problems that touch upon structure sense. In addition to this, Alcumus has tools that can be useful to sustain the kind of meaningful practice required to develop structure sense and other math skills. For example, gamification components like avatars, leaderboards, and adaptive engines deliver new problems based on the performance of previously solved problems. According to the creators, Alcumus is a collection of creative problems with very well-written solutions. In addition, the system can decide if the user needs to keep solving certain types of problems or move on to other topics.

Design Principles for Supporting Structure Sense

Above, we reviewed how currently available algebra learning technologies might be used for supporting structure sense. We draw from this review to identify some design principles applied to design technologies from a structural perspective. Of course, there might be others, but we emphasise these because we possess hands-on experience of their implementation in the MEx platform.

Adaptivity

We agree with Sfard (1991) that learners can conceive mathematical concepts in two complementary ways: structurally, as objects, and operationally, as processes. Typical algebra curricula in secondary and high school tend to emphasise operational notions. Therefore, for most students, the operational conceptions of algebra are necessarily the first step in acquiring new algebraic notions. The transition from operational to structural approaches with symbolic algebra is inherently lengthy and challenging. The pioneering studies of Hoch and Dreyfus (2005, 2006) suggest that, at any given point, learners might possess a "level" of structure sense. However, we argue that the development of structure sense is not necessarily linear. That is, learners might skip levels or show regressions across the levels proposed by Hoch and Dreyfus. Every learner will follow a particular learning trajectory, and trajectories might be hugely different. Therefore, in an ideal world, structure sense instruction should be individualised. This approach is unrealistic in traditional classrooms because teachers need unaffordable time and resources (Dowker, 2017). However, individualised instruction is realistic with adaptive technologies.

Educational platforms with adaptive technologies tailor their content to meet individual needs, thereby implementing individualised learning or differentiated learning. For example, Khan Academy and Duolingo have massively used learning platforms that implement adaptivity. The idea of designing learning technologies with adaptive capacities started to emerge in the early 1980s in the form of student modelling in Intelligent Tutoring Systems (Sleeman & Brown, 1982). One goal of adaptivity is sustaining an optimal learning state, with the premise of delivering content with an adequate degree of challenge. If too easy, the learner gets bored, and if too complicated, the learner gets frustrated. Predicting the next point in a learning trajectory is the "holy grail" of adaptive systems. For structure sense learning technologies, accurate prediction means predicting an adequate task for a learner.

Gamification

Gamification is an approach to design that consists of adapting certain principles and design features developed for computer games to other technological contexts. In learning technologies, a gamified system aims to prompt and sustain a state of optimal engagement over time, that is, a flow state (Hamari & Koivisto, 2014). Some gamification principles for learning technologies include leaderboards for the learner to accumulate points and badges by spending time in the system, for example, solving problems. Platforms like Alcumus and Khan Academy give badges, points, and auditory or visual rewards every time a learner completes an exercise.

Another gamification principle for learning technologies is levelling-up. This assessment process keeps users at a level until mastering the content. For many users, levelling-up feels natural because they are used to content delivered in a sequence of increasing levels of difficulty. Levelling-up helps the user get familiar with the mechanics of a system so that explicit instruction can be minimal. For learning purposes, levelling-up can support processes such as formative assessment. Learners can use what they learn in basic levels for solving more complex exercises in later, more difficult levels.

Variation theory can be beneficial for levelling-up mathematical tasks. According to Watson and Mason (2006), variation can help students to develop a skill. In a sequence of tasks, if designers keep some features while varying others, students can conjecture what is changing and what is keeping the same. Noticing generality by thinking about contrast and similarity can help students appreciate the underlying structures of mathematical objects. Kirshner and Awtry (2004) studied the visual aspect of algebraic expressions and found that the visual similarity of some elements can lead to errors in the understanding of algebra rules. Later in this chapter, we will explain how we used variation and visual salience to design the tasks presented in Mex.

Worked Examples

Khan Academy presents some worked examples in video or text only for explaining solutions. Duolingo uses worked examples with a modelling approach, presenting either the solution to an item or a probable solution. In Alcumus, when learners solve a problem incorrectly, the system invites them to make corrections and try again. If the learner solves the problem correctly, the system outlines the solution, helping learners to achieve a deeper understanding of the problem domain and learn alternative techniques for solving it. Booth et al. (2013) found that presenting both correct and incorrect worked examples may be beneficial for fostering conceptual understanding. Furthermore, their study showed that discussing incorrect examples helps to weaken faulty knowledge and force students to detect and correct errors. Similarly, Star et al. (2015) remarked that having students compare and contrast worked examples can increase students' knowledge.

Interaction Design

Intuitive interaction is a highly desirable property in learning technologies. There are various approaches for designing intuitive ways of interacting with a system, including gestures (e.g., drag-and-drop, shaking), multiple representations, mechanical restrictions, and affordances. The game mechanics and representations of Dragon Box are good examples of the learning potential of intuitive designs. This game helps students learn about equivalence, variables, and commutativity. Instead of introducing these notions explicitly, Dragon Box implements what Habgood and Ainsworth (2011) describe as an intrinsic integration approach because the game's core mechanics integrate the learning content. In this case, the mechanics are structurally equivalent to algebraic manipulations. For example, Dragon Box presents a board split in two and cards with numbers and variables represented with a question mark. The constraints of the interface will make students equate the board by adding equivalent elements on each side. Dragon Box also has visual keys that complement its algebraic mechanics. The system highlights both a similar card and a hole in the left side. These visual highlights act like affordances that invite the player to move a card to the left side of the board and complete the equation.

Above, we mentioned that in Graspable Math, a student could not apply a formula to an incompatible expression. This restriction forces the student to think of an equivalent expression to make its structure more evident. For example, a teacher can scaffold students with a task that presents an expression like $2x + 5$, an instance of the structure $ax + b$, and then let the student modify the numerical values with spinners to see the effect graph. In the next task, the expression might present open texts instead of fixed values, forcing the student to think about structure and which values to assign.

Learning Analytics Dashboards

Feedback is essential in learning systems. Interaction design features like affordances, representations, and mechanical constraints also facilitate learning by giving immediate and implicit feedback. However, explicit feedback is also necessary. Learning analytics dashboards (LADs) are tools that can give sophisticated feedback. LADs can give feedback about social or individual learning processes. Feedback tools that support social learning processes are often termed awareness tools and let learners see information about other users. Awareness tools support interactions that are impossible without digital technologies. For example, educational dashboards that visually display learners' progress can help teachers be aware of the dynamics of a group at the level of individuals. This information can be helpful to orchestrate a classroom, virtually or face-to-face. Achieving this kind of awareness in large classrooms is just impossible without technology.

In learner-oriented environments, LADs can be useful for individuals to be self-aware of their progress. Most current LADs target performance visualisations for informing a learner whether he or she is doing well or poorly, content completion, time spent, or progress compared to other learners or teachers' expectations. This kind of performance-oriented awareness is helpful but limited to support the learner's motivation and engagement. However, LADs can do much more on the way of meeting the general principles of feedback outlined by Sadler (1989): (i) clarifying what good performance is, (ii) facilitate self-assessment, and (iii) offer opportunities for closing the gap between current and good performance. According to Sedrakyan et al. (2020), LADs give cognitive feedback in different forms, for example, corrective feedback about the adequacy of a learner's work, epistemic feedback that stimulates explanations, and suggestive feedback telling the learner how to proceed. We argue that the development of structure sense requires meaningful practice over long periods that the learner can achieve only with constant motivation. Feedback tools like LADs can help students to keep motivated by helping them to keep track of their learning progresses. Already mentioned apps like Khan Academy and Alcumus have LADs for teachers and students. Similarly, Desmos, Geogebra Live, Mathigon, and Graspable Math have live LADs for teachers. Students can connect to a live session created by the teacher to see real-time students' responses and individual work.

Designing Interactions with Algebraic Foundations: The MEx Platform

The MEx platform (http://mexalgebra.xyz/) specifically supports the development of structure sense. The design of MEx engages the user in the solution of a vast repertoire of increasingly complex tasks of algebraic substitution. The tasks help students learn about structure and syntax with mechanics that rely on the function machine metaphor. This metaphor is straightforward: when a student puts an object in the input plate, the machine will take it inside, process it, and transform it into another object, which comes out from the machine. The function machines in MEx have a generating expression (GE). The machine gets one or more input expressions (IE) into its funnels, substitute them with a GE, and transform it into an output expression (OE). MEx has three types of tasks: find output, find generating expression, and find input. The key idea is that given two of these components (IE, OE, or GE), the third needs to be found (Muñoz & Rojano, 2014). MEx is an innovative tool that integrates algebraic foundations and design principles applied to the development of structure sense.

Algebraic Foundations of MEx

The design of MEx started with an analytical process that sought to ensure that each algebraic expression could offer opportunities to work with different aspects of structure sense. The team of researchers that prepared the mathematical content of MEx integrated three algebraic aspects that are crucial to developing structure sense: equivalence, substitution, and process-object duality. Next, we describe the foundations of substitution and the process-object duality. Kieran thoroughly covers algebraic equivalence and its importance for structure sense in this volume (Chapter 3).

Substitution of Algebraic Expressions

The mechanics of the function-machine tasks intrinsically integrate the process of algebraic substitution. The principle of algebraic substitution lies at the core of structure sense. In an algebraic expression, if two symbols are equivalent, then one symbol can replace the other. For example, a student can replace a variable with an expression or vice versa (Freudenthal, 1986). To develop substitution skills and, consequently, structure sense, students need to learn how to recognise several well-known algebraic identities integrated into the expressions of MEx. Some of these identities include $k(a+b) = ka + kb$, $(a+b)^2 = a^2 + 2ab + b^2$, $(a+b)(a-b) = a^2 - b^2$, $(x+a)(x+b) = x^2 + (a+b)x + ab$, and the laws of exponents like $(a^m)^n = a^{mn}$, etc. Students also need to learn how to recognise factors of a number, for example, recognise that 24 is equivalent to 8×3 or that 8 is equivalent to 2^3. Rewriting expressions can be seen as a form of substitution, for example, seeing 2^8 as the square $(2^4)^2$ can be helpful in transformational algebra because to notice that $(2^4)^2$ is equivalent to 2^8, students need to emphasise the square part of the expressions. Fluency with factorisation is also essential because students can use this skill for rewriting expressions.

MEx emphasises the acquisition of algebraic substitution skills because typical mathematics curricula do not offer enough opportunities for their development. Students often learn that algebraic substitution is a numerical substitution. For example, students know how to substitute literals with specific numbers or constants but very often do not know how to substitute algebraic expressions. Solving equations is probably the only mathematical context in which students have opportunities for practising algebraic substitution. The lack of algebraic substitution skills generates difficulties for developing structure sense. Freudenthal (1986) analysed a prototypical error of students who struggle with algebraic substitution. If in the expression $-b$, b is substituted with $b + d$, then the result has to be $-(b + d)$ and not $-b + d$, which is the result obtained by many novice students. To avoid errors, Freudenthal suggests that students should use parentheses. However, he also acknowledges that doing so can be very tedious. Freudenthal goes further and recommends asking students to make relational diagrams so they can think about structure and relations in the expressions. Algebra Arrows implements this idea by helping students explore and learn algebra with binary trees (http://www.fi.uu.nl/wisweb/en/home/welcome.html).

The Process-Object Duality of Mathematical Expressions

The algebraic expressions of MEx should facilitate the transition from a process perspective to an object perspective. Students learn about decomposing expressions and subexpressions by switching between process and object perspectives. Authors like Sfard and Linchevski (1994) and Dubinsky (1991), to name a few, have argued that algebraic expressions have a dual nature, being both processes and objects. One of the central arguments of the process-object duality is that algebraic symbols cannot talk by themselves. A student sees on symbolic expressions what they are prepared to perceive. For example, Sfard and Linchevski (1994) ask the question: when you see $3(x + 5) + 1$, what do you see? They argue that the answer can vary much. Some students might see a computational process, and others might see a number, a function, or a family of functions. Some students might even see a meaningless string of symbols.

Indeed, there can be different ways of seeing an expression. However, we agree with Sfard that novice students cannot achieve the object perspective without mastering the process perspective. One of the critical developments in the transition from process perspectives to object perspectives is learning to think of the chunks that compose expression as objects. Students who achieve this are prepared to find clues for solving increasingly complex problems. A widely known example is the one proposed by Wenger (1987): think about the problem *Solve for v : $v\sqrt{u} = 1 + 2v\sqrt{u+1}$*. At first, one can think of the expression as a two-variable equation. However, by focusing on the *solve for v* part of the instructions, a student can see the expression as an equation with one variable where u is a constant value. From this perspective, the student can recognise that $\sqrt{u}$ and $\sqrt{u+1}$ are numbers.

Design Principles of MEx

Practice is essential for developing structure sense. We have explained that the algebraic expressions of MEx activate structural ways of thinking by emphasising aspects like equivalence, substitution, and the process-object duality. This activation can only be effective if students work with many of these expressions over extended periods. Structure sense is a "naturally occurring" skill among mathematically gifted students (Krutetskii, 1976; Rojano, Chapter 1 in this volume). However, in regular classrooms, students are expected to require substantial amounts of practice to achieve the levels of structure sense required to access higher-education mathematics. MEx supports "meaningful practice". By meaningful practice, we mean an effortful and reflective engagement with tasks explicitly designed to support structure sense.

We should explain how the meaningful practice supported by MEx is different from making students go through hundreds of repetitive tasks. Let us borrow from Hewitt (1996) to make this differentiation. First, students do not need to possess a good understanding of structure to engage with substituting expressions in the function machines of MEx. Below we will see how the function machine metaphor makes the tasks of MEx easily accessible to almost any student with basic arithmetic skills. Second, the focus of attention is not on what students will be practising but on the results of their practising. The practising of structure sense is subordinated to the task of substituting expressions in the function machines. Third, when engaged in the task, the student does not need a teacher to see whether she is correct or not. The consequences of her decisions in the function machines can help her learn about structure.

There are potential drawbacks associated with extended periods of practice in mathematics, such as potential boredom and the fixation with mechanistic operations. MEx addresses these issues with the implementation of design principles like levelling-up and gamification. Also, MEx can reach a broad audience that may not be fluent in algebra. Therefore, MEx includes numerical tasks similar to "guess my rule" tasks. In these tasks, the only literals involved will be those shown in the process, that is, the ones that describe the GE. Next, we turn to describe the design principles of MEx.

Representational Metaphor

The function machine metaphor is straightforward and representationally powerful. We argue that most users of MEx will be familiar with this metaphor because input-output mechanisms are pervasive. For example, students can see vending machines of all kinds as input-output machines. In terms of representational features, the function machine metaphor embodies the object-process duality of mathematical conceptions (Tall et al., 2000). In MEx, users can think about algebraic expressions as objects. They can also reflect on the substitution process that transforms the input expression into the output expression.

It is common to find function machines that receive numbers and "spit-out" numbers, for example, in "Guess my rule" tasks that introduce the function concept (Carraher & Earnest, 2003). However, most of the function machines of MEx are not numerical and, instead, the domain and codomain are algebraic expressions. In Figures 4.2a and 4.2b, we compare a text-based equivalent to the visual version of the input-output machine

a)

$(4x^2 - 8x + 6)(8x^2 + ax - 8)$

$32x^4 - 56x^3 + 76x - 48$

b)

Find a that satisfies
the following
equation:

$(4x^2 - 8x + 6)(8x^2 + ax - 8)$
$= 32x^4 - 56x^3 + 76x - 48$

c)

$a(x^2 + 1)$

$(x^2 + 1) + (x^2 + 1) + 2(x^2 + 1)$

d)

$a(x^2 + 1)$

$x^2 + 1 + x^2 + 1 + 2(x^2 + 1)$

e)

ab

$(x^2 + 1) + (x^2 + 1) + 2(x^2 + 1)$

f)

ab

$x^2 + 1 + x^2 + 1 + 2(x^2 + 1)$

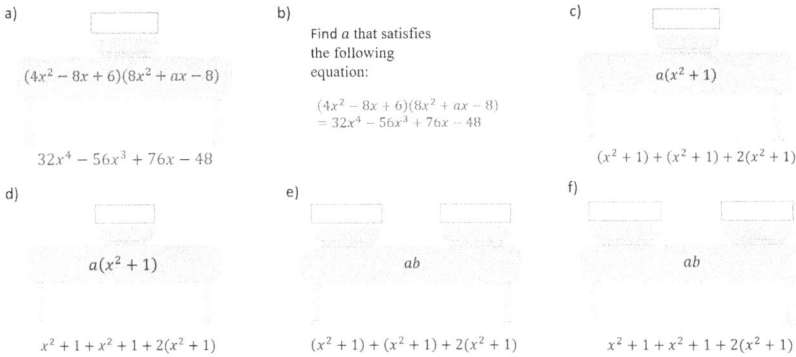

Figure 4.2 (a) An advanced task in MEx; in (b), task (a) without the meta-phor machine. Some variations of similar tasks: (c) brackets drawing attention to subexpressions $x^2 + 1$; (d) brackets are removed from the first two terms of c) decreasing in that way the visual salience towards $x^2 + 1$; (e) brackets are added, but the structure in the gener-ative expression is a product whereas the output expression is a sum of three terms; (f) no visual salience and different structures between generative expression and output expressions.

metaphor. We argue that because of the visual metaphor, in Figure 4.2a the users can understand what they need to do without words or with very few words.

Levelling-up

Another feature has to do with the so-called levelling-up in serious games that we mentioned before. MEx organises the tasks in order of difficulty, so the experience that students gain in previous exercises helps them solve forthcoming exercises. In the current version, the pathways pre-sented to the user are pre-defined. However, an adaptive model based on cognitive features will be available in future. An exercise recommender system module in MEx considers previous answers to propose the next task, which has a reasonable probability for the user to answer correctly but still be a challenge. There are easy tasks in MEx. However, other tasks requiring algebraic transformations that students cannot solve with non-structural or trial and error strategies, like the one in Figure 4.2a. Nevertheless, the user can decide to skip some problems and try some others without completing them, fostering student agency.

Symbolic Computation CAS as Symbolic Engine
and Substitution, Equivalency, and Randomness

MEx uses a CAS called Giac to make all symbolic computations includ-ing comparing expressions to determine if they are equivalent, simplifica-tion, and the most important, substitution (Parisse & De Graeve, 2008).

MEx also has an item-generation engine based on random coefficients. This engine identifies if the user tries to solve an item repeatedly and presents a slightly different instance each time.

Thanks to the CAS Giac, the task designers could generate interesting expressions using the substitution and some other transformations. Some of those expressions have a deeper parse tree associated. Although the student does not see the parse tree, task designers use it as a metric for generating more complex tasks.

For example, with a base structure like $a^2 + 2ab + b^2$, it is straightforward using the Giac substitution to generate other expressions like (a) $(x+2)^2 + 2(x+2) + 1$, (b) $(x+2)^2 + 2(x+2)(x+3) + (x+3)^2$; or (c) $(x+1)^2 + 2(x+1)(x-1) + (x-1)^2$. Moreover, one can present those expressions in an expanded way or in a factored way. MEx also welcomes all the possible equivalent expressions, even if there is a naturally defined correct expression for a given task. For example, if the expected answer is $2x + 2y$, there is a parser checking if this is equivalent to $2(x+y)$. Also, the newest version shows a visual step-by-step solution to the task.

Visual Salience

With the CAS module, MEx task designers control how an expression appears to the user. For example, one task can appear to the user in an expanded form like $2x + 3 + 5x + 6$. This feature is convenient to show different equivalent expressions in various forms, with or without brackets, simplified or expanded to increase or decrease the visual salience that we commented on in the previous section. Having control of how an expression appears is also helpful to get the user to know different notations. For example, sometimes, the expressions are written using a dot, brackets, or just concatenation to denote multiplication: $2x, 2 \cdot x, (2)(x), 2(x)$.

The designed tasks take advantage of the **visual salience** of algebraic expressions and the associated parsing tree to decrease or increase the difficulty of the tasks. One way to increase visual salience is through brackets, as studied by Hoch and Dreyfus (2005).

For example, if students are asked to simplify $(x^2 + 1) + (x^2 + 1) + 2(x^2 + 1)$, brackets serve as a visual clue to draw attention to the subexpression $x^2 + 1$. There is a possibility that students think of $x^2 + 1$ as a single entity and easily add all those terms to see the whole expression as $4(x^2 + 1)$. In contrast, if MEx presents the same expression without the first two brackets, like in $x^2 + 1 + x^2 + 1 + 2(x^2 + 1)$, non-fluent students will likely simplify like terms first to obtain $2x^2 + 2 + 2(x^2 + 1)$, then remove parenthesis to get $2x^2 + 2 + 2x^2 + 2 = 4x^2 + 4$. In Figure 4.2, we show some of those variations with similar tasks. In summary, using the machine metaphor and the visual salience, task designers can make some variations to simplify or increase the difficulty.

Gamification

In the first usability tests, students asked for gamified features. They asked to get visual or auditory feedback when they succeed. They also want to see their names on screen and be able to personalise their accounts. Some of them asked for an avatar and to be able to customise it. In that sense, we believe that young students are getting used to the features presented in games or gamified math platforms like Khan Academy. They may even expect some of those gamified features as the norm.

Conclusion

In this chapter, we reviewed a range of algebra learning tools, including MEx. Our emphasis was on describing the potential of different technological tools for supporting the development of structure sense. This approach complements current views of algebra learning technologies, which emphasise classroom experiences with technologies in which teachers must play active participation. We also reviewed some teacher-oriented digital tools and showed practical uses to mediate structure sense in algebra. Additionally, we describe some design principles to create learner-oriented tools to the specific goal of promoting structure sense. In this regard, we presented the learning environment MEx with a design perspective.

Being aware that practice plays an important role in developing algebraic proficiency and structure sense, we emphasised that we are not referring to a mechanical view of drill and practice, but rather to purposeful practice and mathematical reflection in solving algebra tasks. We believe that the object-process perspective is key to promoting this mindful and reflective practice. As it is known, most algebra novices see expressions as processes, and tools like MEx can be helpful to promote the shift towards the object-perspective. Moreover, once an expression or subexpression is seen as an object, students can switch between process and object at will to solve a challenge in MEx. Additionally, level design often used in gamification and serious games can play a key role in fostering the development of structure sense, and research about variation theory may be key to this design.

As of the review of tools and design principles, we conclude that more research needs to be done on topics about structure sense with and without technology. In the case of technology usage, there are still open questions, for example: Can the reviewed tools foster structure sense? What features does an algebraic task need to have to promote a purposeful practice? What digital tools can promote structural thinking in algebra and how?

The next chapter will focus on learners' interaction using MEx and present interesting results about some of these questions.

References

Booth, J., Lange, K., Koedinger, K., & Newton, K. (2013). Example problems that improve student learning in algebra: Differentiating between correct and incorrect examples. *Learning and Instruction, 25*, 24–34.

Caglayan, G. (2014). Visualising number sequences: Secondary preservice mathematics teachers' constructions of figurate numbers using magnetic colour cubes. *The Journal of Mathematical Behavior, 35*, 110–128. https://doi.org/10.1016/j.jmathb.2014.06.004

Carraher, D. W., & Earnest, D. (2003). Guess my rule revisited. *International Group for the Psychology of Mathematics Education, 2*, 173–180.

Crook, C. (2001). The social character of knowing and learning: Implications of cultural psychology for educational technology. *Journal of Information Technology for Teacher Education, 10*(1–2), 19–36. doi: https://doi.org/10.1080/14759390100200100.

Dowker, A. (2017). Interventions for primary school children with difficulties in mathematics. *Advances in Child Development and Behaviour, 53*, 255–287. https://doi.org/10.1016/bs.acdb.2017.04.004

Dubinsky E. (1991). Constructive aspects of reflective abstraction in advanced mathematics. In: Steffe L.P. (eds) Epistemological foundations of mathematical experience. Recent research in psychology. Springer, New york, NY. https://doi.org/10.1007/978-1-4612-3178-3_9

Freudenthal, H. (1986). *Didactical phenomenology of mathematical structures* (Vol. 1). Springer Science & Business Media.

Habgood, J., & Ainsworth, S. E. (2011). Motivating children to learn effectively: Exploring the value of intrinsic integration in educational games. *The Journal of the Learning Sciences, 20*(2), 196–206

Hamari, J., & Koivisto, J. (2014). Measuring flow in gamification: Dispositional flow scale-2. *Computers in Human Behavior, 40*, 133–143.

Hewitt, D. (1996). Mathematical fluency: The nature of practice and the role of subordination. *For the Learning of Mathematics, 16*(2), 28–35

Hoch, M., & Dreyfus, T. (2005). *Structure sense in high school algebra: The effect of brackets.* International Group for the Psychology of Mathematics Education.

Hoch, M., & Dreyfus, T. (2006). Structure sense versus manipulation skills: An unexpected result. *Proceedings of the 30th Conference of the International Group for the Psychology of Mathematics Education, 3*, 305–312.

Jupri, A., & Sispiyati, R. (2020). Students' algebraic proficiency from the perspective of symbol sense. *Indonesian Journal of Science and Technology, 5*(1), 86–94.

Kieran, C. (2018). Seeking, using, and expressing structure in numbers and numerical operations: A fundamental path to developing early algebraic thinking. In C. Kieran (Ed.), *Teaching and learning algebraic thinking with 5- to 12-year-olds: The global evolution of an emerging field of research and practice.* Springer.

Kirshner, D., & Awtry, T. (2004). Visual Salience of algebraic transformations. *Journal for Research in Mathematics Education, 35*(4), 224–257. DOI: https://doi.org/10.2307/30034809

Krutetskii, V. A. (1976). *The psychology of mathematical abilities in schoolchildren.* University of Chicago Press.

Muñoz, V., & Rojano, T. (2014). Algebraic expression machine: A web ad hoc learning environment for developing structure sense. *Proceedings of the 3rd International Constructionism Conference*, 540–541.

Parisse, B., & De Graeve, R. (2008). *Giac/Xcas, a free computer algebra system. Technical report.* University of Grenoble.

Sadler, D. R. (1989). Formative assessment and the design of instructional systems. *Instructional Science*, 18(2), 119–144.

Sedrakyan, G., Malmberg, J., Verbert, K., Järvelä, S., & Kirschner, P. A. (2020). Linking learning behavior analytics and learning science concepts: Designing a learning analytics dashboard for feedback to support learning regulation. *Computers in Human Behavior*, 107. https://doi.org/10.1016/j.chb.2018.05.004

Sfard, A. (1991). On the dual nature of mathematical conceptions: Reflections on processes and objects as different sides of the same coin. *Educational Studies in Mathematics*, 22(1), 1–36. doi: https://doi.org/10.1007/BF00302715.

Sfard, A., & Linchevski, L. (1994). The gains and the pitfalls of reification – the case of algebra. In *Learning mathematics* (pp. 87–124). Springer.

Sleeman, D., & Brown, J. S. (1982). *Intelligent tutoring systems.* Academic Press.

Star, J. R., Pollack, C., Durkin, K., Rittle-Johnson, B., Lynch, K., Newton, K., & Gogolen, C. (2015). Learning from comparison in algebra. *Contemporary Educational Psychology*, 40, 41–54.

Tall, D., McGowen, M., & DeMarois, P. (2000). The Function Machine as a Cognitive Root for the Function Concept, *Proceedings of the Annual Meeting of the North American Chapter*, Vol. 1, 255–261.

Watson, A., & Mason, J. (2006). Seeing an exercise as a single mathematical object: Using variation to structure sense-making. *Mathematical Thinking and Learning*, 8(2), 91–111.

Weitnauer, E., Landy, D., & Ottmar, E. (2017). Graspable math: Towards dynamic algebra notations that support learners better than paper. *FTC 2016 – Proceedings of Future Technologies Conference*, 406–414. https://doi.org/10.1109/FTC.2016.7821641

Wenger, R. H. (1987). Cognitive science and algebra learning. In A. H. Schoenfeld (Ed.), *Cognitive science and mathematics education* (pp. 217–251)). Lawrence Erlbaum Associates.

Xolocotzin, U., & Pretelín-Ricárdez, A. (2015). Exploring the historical development of computer games research in mathematics education. *Proceedings of the 12th International Conference on Technology in Mathematics Teaching ICTMT 12.*

5 Transformational Algebra, Structure Sense, and Notes on a Semiotic Interpretation

Armando Solares-Rojas and Teresa Rojano

A Semiotic Explanation of the Development of ASS

Since the 1980s, when the teaching of algebra that was based on rules leading to a blind manipulation of symbol chains was questioned, the meaning construction of symbolic algebra objects has been considered a fundamental piece of learning algebraic syntax. To try to explain the processes that take place in this meaning construction, in this chapter we adopt a semiotic perspective in which algebraic expressions and related tasks are considered *texts*, and *signification* is the result of *sense production* in respect of these texts. The theoretical perspective in question is based on the ideas developed by Peirce (1982) and which, together with other authors, we have used in previous works to address topics such as learning to solve equations (Filloy et al., 2008) and systems of linear equations (Filloy et al., 2003) and learning algebraic methods of solving word problems (Filloy et al., 2010).

Specifically, our interest is to describe how the use of structure sense that occurs when subjects solve transformative algebra tasks can be interpreted in terms of signification processes associated with the algebraic expressions themselves and their transformations.

The opening chapter proposes that by working with activities involving the symbolic manipulation of algebraic expressions in a learning environment with leveling-up design it is feasible for subjects with different mathematics backgrounds to mobilize their knowledge when they face activities that require developing and implementing ASS capabilities. Data obtained in the study carried out with the expression machine (MEx) (described in Chapter 4) show that when working in this web environment individuals with different types and degrees of mathematical background are able to solve transformative algebra tasks that require an increasingly complex use of ASS. To illustrate the above, this chapter presents the cases of two users with dissimilar mathematical experiences and whose analysis is based on identification of paths drawn in their interaction with MEx, as well as actions that show indications of the use of ASS elements. In a later section, we interpret the ASS capabilities

DOI: 10.4324/9781003197867-5

required by MEx tasks in terms of the signification processes that under-lie the transformative algebra activity involved in solving those tasks.

Development of ASS, Leveling-Up Design and Task Types

Chapter 4 describes the leveling-up design of algebraic tasks that MEx proposes to users. The levels correspond to the complexity of the expressions involved in the tasks, which has to do with both the compound terms that make up the expressions (parse tree) and the presence (or lack thereof) of markers or elements of visual salience[1]. Depending on these two traits, the solution to each task requires one or more of the capabilities considered by Hoch in his definition of ASS, that is, identification and application of a rule in its simplest form; application of a rule in a complex expression, and treatment of a compound term as though it were an entity (Hoch, 2007). These capabilities depend largely on the use of numerical or algebraic substitution.

But in addition to the degree of complexity (understood in the above-mentioned terms), the three types of tasks included in MEx (described by Muñoz & Xolocotzin in this volume) may require other capabilities. For example, type I tasks consist of: given two or more expressions as inputs and a generating expression, respondents are asked to find the expression that results from applying the generator to the inputs, which involves a direct action. The task in Figure 5.1a deals with applying the generating function "add inputs 49-1 and 49 + 1 of dishes a and b". The answer is written in the lower portion of the machine.

Whereas in type II tasks, the generating expression and the resulting expression (output) are given, and respondents are asked to find the inputs (see Figure 5.1b). In the latter case, the task does not necessarily involve applying a rule to the given elements, but rather the starting point is the result of that application, which is an algebraic or numerical expression, and which may involve breaking it down, identifying terms, operations and rules of origin. Depending on the complexity of the expression (algebraic or numerical), this type of task may involve breaking down such expression in one or several ways and undertaking a process of testing

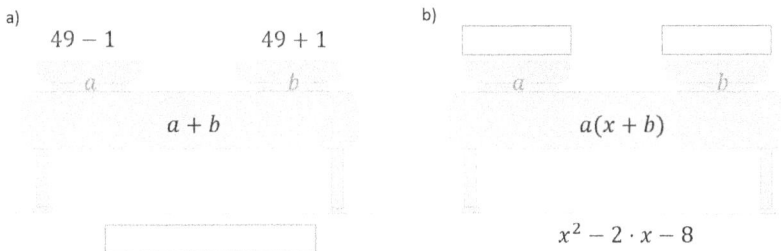

Figure 5.1 (a) An example of type 1 task, (b) a type 2 task.

and refinement, applying the generating rule to the parts resulting from break down. Or as in the example in Figure 5.1b, the generating expression $a\,(x + b)$ may be suggestive of the type of break down/factoring ("$x + b$ times something") of the expression $x^2 - 2x - 8$. Another way to address type II tasks is by solving (for input(s) a and b) the equation that results from equating the generating expression to the output expression.

Type III tasks, in which the input(s) and output are given and respondents are asked to find the generating expression, may require the combination of different actions, such as applying hypothetical rules or generating expressions to input elements and verifying whether the output expression results from application of any of those rules. However, the presence of visual salience elements can simplify identification of the generating rule. For example, one can see in the task in Figure 5.2a that the visual salience trait (+83 in a and −83 in b) evokes a possible elimination of additive reverse terms, leading to the (correct) generating expression $a + b$, which is an algebraic sum. Similarly, the task in Figure 5.2b has the numerical traits of visual salience −7 in a, −9 in b, and 63 in the output tray, which evokes the (correct) ab rule, which is an algebraic product (rule of signs) and which is consistent with the product of the terms of the task, which include variable x. However, in the absence of visual salience traits, type III tasks may represent great difficulty and require a more complex structural analysis. See, for example, the task in Figure 5.2c which

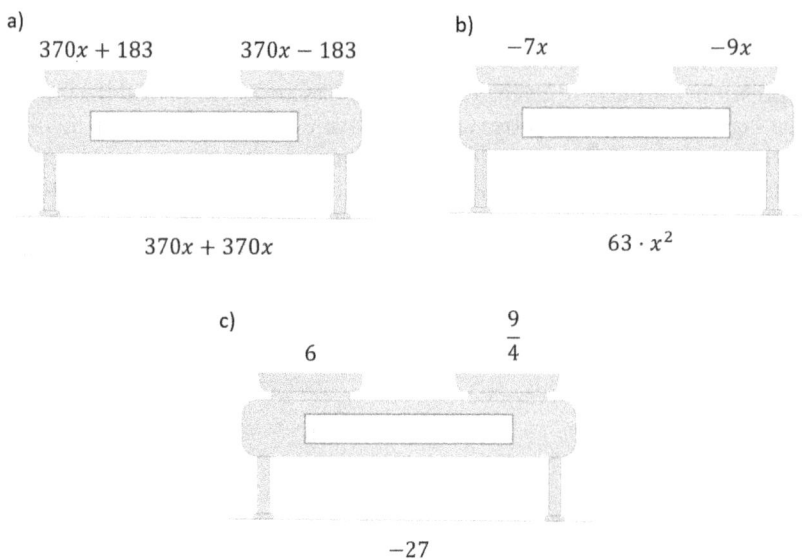

a)

$370x + 183$ $370x - 183$

$370x + 370x$

b)

$-7x$ $-9x$

$63 \cdot x^2$

c)

6 $\dfrac{9}{4}$

-27

Figure 5.2 Different visual salience traits in type III tasks: (a) the visual salience evokes a possible elimination of additive reverse terms, (b) the numerical traits of visual evokes the (correct) ab rule, (c) absence of visual salience traits.

is arithmetic in nature and requires numerical structural capabilities, in this case, referring to fractions.

Strategies for solving type III tasks can be diverse; however, they all involve processes of comparing the input expression or expressions with the output expression, where the notion of algebraic (or numerical) equivalence underlies the process.

Unlike exercises traditionally included in algebra texts that use key-words such as "factoring, developing or simplifying an expression", transformative algebra actions required by most MEx type II and III tasks involve breaking down and recomposing expressions, reversibil-ity processes, testing, and refinement, as well as the use of a repertoire of known rules or the formulation of compound rules (elaboration of the generating rule), as well as the implicit use of fundamental con-cepts, such as algebraic substitution and equivalence or the notion of equation.

The topics or category of tasks included in MEx are labeled as follows:

1-10, 1-100, 1-1000 are tasks of the arithmetic type that include addition, subtraction, multiplication, and division with numbers; Algebra, with subtopics: factoring, multiplication, and notable products; Exponents and Roots, which includes numerical cases and algebraic expressions; Algebraic Expressions, which consists of tasks of operations with algebraic expressions; Fractions 1 and 2, which include operations with fractions; Hierarchy of Operations; Negatives; and Challenges, which includes Advanced 1 and Advanced 2 sections.

In turn, in each category the tasks are classified according to com-plexity levels: basic, intermediate, high and, as already mentioned, in the Challenges section there are two levels, advanced 1 and advanced 2.

It is worth mentioning that from the outset in arithmetic tasks work with MEx activities requires knowledge of the basic structural proper-ties of integers and fractional numbers and their operations. While to solve algebraic tasks, knowledge of the basic structural properties of algebraic expressions, as well as the rules for transforming and operat-ing them with each other are required. This separation into arithmetic tasks and algebraic tasks allows for use of MEx by users who have not yet had any algebra instruction.

Development of ASS: The Study with the Expression Machine

The study carried out with the MEx within the framework of the pro-ject entitled "Developing algebra structure sense in a digital interactive environment with an adaptive feedback system"[2] has the general pur-pose of proving the feasibility to foster development of high algebraic competences in heterogenous groups of students, through the use of a technology environment that promotes development of ASS. To achieve this purpose, two specific objectives were set:

- To design, develop, and test a web environment with an adaptive system to give feedback to users, when they engage in transformative algebra tasks.
- To probe the notion of algebra structure sense, departing from the Hoch and Dreyfus definition, through the analysis of the deep structure of algebraic expressions (parsing), and taking into account outcomes from the experimental work.

The following hypotheses guided development of the project:

- Based on an extension of the notion of algebra structure sense it is possible to design a web learning environment with a feedback system according to the user level of algebraic competence.
- A frequent and intensive interaction with the web environment allows students to reach high levels of algebraic competence, due to the feedback type and to the activity types designed according to the structural complexity of the algebraic expressions (internal structure).
- Over the course of 2 weeks, a group of 10 participants enrolled in individual work with MEx, lasting 4 to 6 hours on average. During this stage, each participant freely traveled a different route, the trail of which in part reveals their initial choice of subject and level of tasks and in part their performance in tasks of varying degrees of complexity.

In general, interaction with MEx consists of the following: once the user chooses a topic or category (e.g., "Integers") and level (for example, "Intermediate"), the system proposes a task that corresponds to that topic and level and the user types in her response that will be the "output", "input", or "generator expression", depending on whether it is a type I, II, or III task. The data that are recorded in the MEx system for each user and that are used in the analysis consist of: (1) the set of tasks solved; (2) the order in which they were addressed; (3) the response for each task; (4) the time spent per task; (5) the number of attempts per task; (6) the attempts made on each task.

Each answer may be the product of a quick user perception of the structure of expression or expressions involved in the task, or the result of a structural analysis that may have been carried out with paper and pencil algebra outside of MEx. Or it may be the result of a combination of those two actions. In most cases, the foregoing can be determined based on the type and number of attempts made and the time spent solving the task. The collection of data from the participants' individual work with MEx was carried out online, without the intervention of any other subject.

In order to illustrate the types of responses to which the leveling-up design of MEx tasks gives rise, the cases of two of the 10 participants with well differentiated mathematical backgrounds are presented below.

Francisco, Lulú, and MEx Tasks

The cases of Francisco and Lulú were chosen based on the criterion that they have mathematical schooling focused on different purposes. Lulú is a former primary school teacher. She gained teacher training at the college and, after some years of working at a primary school, she did graduate studies in mathematics education. For his part, at college Francisco trained as high school mathematics teacher. Then he did graduate studies in mathematics education.

Participant productions are the data source used for the analysis, and they are recorded in the MEx system itself as "attempts" on each of the tasks. Special attention was paid to the tasks in which each participant made two or more attempts, as well as the score, which is either zero (in the case of an incorrect answer) or 10 (in the case of a correct answer). Resolution time for the tasks was also taken into account. When such time is very short it suggests an immediate perception of the structure of the expressions involved and their relations.

The Case of Francisco

Francisco, who will be referred to hereinafter as F, is a high school mathematics teacher with graduate studies in mathematics education. His training is that of a university student who is studying to become a mathematics teacher at different educational levels, which includes courses in abstract algebra and other courses in advanced mathematics. Francisco did a run through of MEx, making his way through all categories. He solved most tasks with a single attempt and some with 2 or 3 attempts, in a time that in most cases varied from 5 to 90 seconds per task. However, in the Challenges category he spent up to 10 minutes on two tasks.

Below is an overview of F's run through the MEx task network. In addition, a set of tasks was chosen in which F made two or more attempts, and in which such attempts suggest the type of strategy used to solve them. His responses and attempts are described by category (topic), including the level in each case.

Performance of F on a Selection of MEx Tasks

In categories 1-10, 1-100, and 1-1000, which correspond to arithmetic topics, F solved the proposed tasks correctly with an average time of between 5 and 10 seconds at each task and in one single attempt, except in two cases, where he made numerical miscalculations that he corrected immediately. In these two cases, the resolution time was approximately half a minute.

In the Fractions 1 and 2 categories, F solved the tasks correctly in an average time of 1/2 to 1 minute in one single attempt, except for task 148 (Fractions 2). In the latter task, respondents are asked to find the generating expression and the task involves division of fractions. F solved it in two attempts (in about half a minute), in the second of which he adjusted his initial response, as can be seen in the following description:

> *Task 148* from "Fractions 2". [Inputs: a: $\frac{5}{2}$; b: $\frac{3}{4}$; generator expression: ?; output: $\frac{20}{3}$. First try: $\frac{a}{b}$; second try (correct): $\frac{2a}{b}$]

In the Algebraic Expressions category, F solves all tasks correctly and in his first try, except for three of them. One of those tasks (intermediate complexity level) is about finding the generating expression, which applied to inputs a: $5x$ and b: $6x$ results in output $-x$. That is, the task is about answering the question: What operation has to be done between a and b, the result of which is $-x$? As already mentioned, in such tasks there is a structure sense requirement that involves formulating a hypothetical proposal of the operation (expression) and testing it, which differs from the direct application of an operation on the input elements. That strategy can be seen in F's (correct) tries $b - a$ and $a - b$.

F successfully solves tasks in the Operations Hierarchy and Algebra-Multiplication categories in a single try and within a time of 0.18 to 0.5 minute and 0.1 to 1.35 minutes, respectively. Except for task 317 in the second of these categories, where respondents are asked to find the input given the generator expression $(a)(y)$ and the output $xy + 4y$, F carries out tries a: $-(-x - 4)$; a: $(x + 4)$; a: $x + 4$ (correct), which indicate a procedure of inspection and comparison of the given expressions and not of solving for a the equation $(a)(y) = xy + 4y$, where the latter strategy leads directly to the response.

F successfully solves Algebra-Polynomials, Notable Products, Common Factor, and Square Difference tasks (all of intermediate complexity), in times between 0.35 minute and 1 minute. For the most part, solving the tasks involves direct application of known rules or algebraic identities or identification of notable products in simple expressions.

As the name suggests, several tasks in the Challenges-Advanced 1 and 2 categories represented real challenges for most participants. F's work should be noted in particular in the following cases:

> *Task 380*, high level. [3 attempts, resolution time 0.78 min. Input: a: ?; generator expression: $2a + x$; output: $3x + 2$. First try $2x + 2$; second try $x + 2$; third try (correct) $x + 1$]

As previously mentioned, in these tasks (type II) the presence of several attempts is indicative that the strategy of inspection and comparison of expressions was used, that is, that ASS capabilities were implemented.

Task 385, high level. [3 tries, resolution time 1.23 minutes. Input: a: ?; generator expression: $a-(2x+3)$; output: $3x-3$. First try $5x+6$; second try $5x-6$; third try $5x+6$]

F makes three failed attempts. This is a task with elements of visual salience, as all one has to do is note that to get $3x$ in the output a must be $5x$ and the rest is a change of sign +3. Moreover, the resolution for a of the equation $a-(2x+3)=3x-3$ is trivial. The three attempts reveal a strategy of inspecting the given expressions, in which F did not successfully treat the numerical portions.

Task 395, high level. [2 tries, resolution time 0.48 minute. Input: a: ?; generator expression: $(6+16x)(8-ax)-(2x+3)$; output: $-32x^2+116x+48$. First try: a: -2; second try (correct): a: 2]

This task can be solved by inspecting the terms whose product is the term in x^2 (i.e., by observing that $(16x)(-ax)=-32x^2$) from which it follows that the input $a=2$. Applying the strategy of solving for a the equation can $(6+16x)(8-ax)-(2x+3)=-32x^2+116x+48$ possibly take longer than half a minute, which was the time spent by F on both attempts. However, in this case it is not possible to infer which of the two strategies he could have used.

Task 440, high level. [2 incorrect tries, resolution time 0.93 minute. Input: a: ?; generator expression: a^2+2x+4; output: x^2-2x+8. First try: a: x^2-2x; second try (correct): a: x^2-2x+1]

The correct answer a: $x-2$ can be obtained by inspecting the expressions, for example, by observing that a^2 must contain the term x^2 a term in x and a constant term, thus a must be a binomial of the form $x\pm una\ constante$, and the value of that constant can be inferred from the constant terms of the generating expression and the output. Another way to proceed is to solve the corresponding equation for a. F's two attempts reveal that his search was made by inspection, but that he focused on finding a as a quadratic expression in x rather than as a linear expression.

Task 441, high level. [2 tries, resolution time 3.8 minutes. Input: a: ?; generator expression: $ax-x$; output: x^3-2x^2. First try: x^2-2x; second try (correct): x^2-2x+1]

From the resolution time (almost 4 minutes), one can infer that F performed paper and pencil algebra activity. However, the difficulty of solving this task by inspecting expressions may be comparable to solving the corresponding equation (which involves performing a division of polynomials), hence the type of strategy used by F cannot be inferred.

> *Task 443*, high level. [2 tries, resolution time 0.68 minute. Input: *a*: ?; generator expression: $a+(a+1)+(a+2)$; output: $3x$. First try: *a*: $a-1$; second try (correct): *a*: $x-1$]

Noticing that the generator expression is the sum of three consecutive terms can help mobilize ASS capabilities to solve this task. However, this strategy and the resolution of the corresponding equation have a similar level of difficulty and could be solved at similar times, hence it is not possible to infer what approach F used.

> *Task 444*, high level. [2 tries, resolution time 1.02 minutes. Input: *a*: ?; generator expression: $(a-1)x$; output: x^2+x^3. First try: $x-x^2+1$; second try (correct): $x+x^2+1$]

This task can be solved quickly by inspection, by factoring in the output $x(x+x^2)$ and matching the expression with $(a-1)x$ or by solving the corresponding equation for *a*. The type of attempts made by F reveal a strategy of inspection and comparison of expressions, since the equation resolution strategy leads directly to a response.

> *Task 386*, high level. [2 attempts, resolution time 0.68 minute. Input: *a*: ?; generator expression: $x+a^2$; output: x^2-x+1. First try: *a*: $(x-1)^2$; second try (correct): *a*: $x-1$]

In this task, the visual salience element enables seeing that the expression x^2-2x+1 (very similar to the output) is factorized as $(x-1)^2$, thus reaching the correct answer *a*: $x-1$. F corrects his response on the second attempt, changing the quadratic expression to a linear expression.

> *Task 387*, high level. [2 attempts, resolution time 1.2 minutes. Input: *a*: ?; generator expression: $3a+5$; output: $5(3x+5)$. First try: *a*: $x-10$; second try (correct): *a*: $5x-10$]

The number and type of attempts made by F indicate a strategy by inspection and comparison of expressions, since the resolution of the corresponding equation leads directly to the response.

> *Task 398*, high level. [2 + 3 attempts, resolution time $10+14$ minutes. Input: *a*: ? ; generator expression: $16a^2+24$; output: $4(x^2+1)$. First attempt *a*: (x^2-5); second attempt *a*: ; third–fifth attempts, failed]

This is a very difficult task if one wants to solve it by mobilizing ASS capabilities. The number and type of attempts made by F are indicative of a strategy of inspection and comparison of expressions, since the resolution strategy for a the corresponding equation $16\,a^2 + 24 = 4(x^2 + 1)$ leads directly to the solution $a = \pm\frac{\sqrt{x^2-5}}{2}$.

> *Task 399*, high level. [0 attempts, resolution time 0. Input: a: ? ; generator expression: $(ax + 5)(-4)$; output: $-28x - 20$]

This task can be difficult to solve by inspecting and comparing of expressions; however, finding the solution is easy by solving equation $(ax + 5)(-4) = 28x - 20$ for a. The correct answer is a: 7. As for F's performance, no attempts or resolution time were recorded. However, he did perform the task and his work was probably entirely done with paper and pencil algebra.

ASS, MEx Tasks, and F's Attempts

F's actions in his run through the categories and levels of MEx tasks reveal skill in the calculation and use of numerical properties, as well as in the application and use of manipulative algebra properties and techniques. These numerical and algebraic competencies are mainly deployed in the responses to tasks focused on operative algebra and arithmetic, like some from categories Fractions, Algebraic Expressions, and Hierarchy of Operations. F's solutions to those tasks constitute baggage that he put into use in addressing the types of tasks that also require an unusual mobilization of capabilities related to the structure of the expressions involved. This is the case with type III tasks (find generator expression) in which it is necessary to produce a hypothetical expression and test whether it meets the conditions of the task. In keeping with the above, the attempts made by F in type III tasks reveal the use of a strategy of inspection and comparison of the given expressions (input and output) to develop the generator expression and then make the necessary adjustments.

The performance of F in type II tasks (find input) shows the use of the same inspection and comparison strategy of the given expressions (in this case generator expression and output), which is reflected in the attempts made. It should be noted that solving the generator expression = output equation for a (input) is a strategy (which is actually a technique) that safely leads to the response in these types of tasks. However, in this category's analyzed tasks F proceeded by inspection and comparison rather than the application of a technique, which denotes mobilization of structure sense capabilities.

In short, although in a couple of tasks F seems to have not taken advantage of visual salience traits to solve them, in most cases analyzed F shows a tendency to use strategies that involve capabilities of perception and/or analysis of the structure of the expressions involved, analysis based on the comparison of those structures with each other. This is observed, for example, in F's attempts on tasks: 140 Fractions 2 (intermediate level); 317 Algebra-Multiplication (high level); 380, 395, 440, 441, 443, and 444 Challenges-Advanced 1; 386 and 398 Challenges-Advanced 2. It is worth mentioning that such a comparison of structures requires – in addition to inspection of the structure of each of the algebraic expression – mathematical thinking that may include logical thinking and reversible processes, among other things. That is, the sense of the algebraic structure that is brought into play in this type of activity goes beyond recognition of a rule and use of algebraic substitution to work with a compound expression or term as if it were an entity. In this regard, both the very design of some of MEx tasks and the strategies used by F in his attempts to solve them further expands the set of capabilities associated to the capability of being aware of the structural features of algebraic objects.

The Case of Lulú

Lulú, whom we will refer from here onward as L, is a primary school teacher with doctoral studies in mathematics education. Her initial training as a teacher is aimed at teaching all subjects at the primary level, from mathematics and language to science and social sciences, with substantial emphasis on the pedagogical aspects and general didactics of teaching primary contents. In her graduate training, Lulú took several courses in arithmetic and algebra, again with emphasis on the didactic aspects of the contents studied.

Lulú ran through all categories and solved most MEx tasks in one or two attempts, although in some cases she made three and even four attempts. In respect of time, the range is broad; from 6 seconds to several exercises that she spent around 10 to 13 minutes on. In fact, she spent hours on a couple of them although this was surely due to having left the session open. For example, she correctly solved exercise 285, corresponding to the category of Operations Hierarchy (intermediate complexity level), in one attempt after 367 minutes (just over 6 hours).

Below is an overview of L's run through of MEx tasks. In addition, a set of tasks was chosen in which L made two or more attempts, and in which these attempts suggest the type of strategy used to solve them. Her responses and attempts are described by category (topic), including the level in each case.

Performance of L on a Selection of MEx Tasks

In categories 1-10, 1-100, and 1-1000, which correspond to topics of arithmetic, L solved the tasks correctly, in general with times between 6 and 60 seconds each. However, for this category the number of attempts per exercise was uniform: two attempts in almost all exercises. Apparently, her errors were due above all to mistakes in numerical calculation.

In the Fractions 1 and 2 categories, L solved almost all tasks correctly, except tasks 120, 122, 141, and 142 for multiplication and division of fractions. The time spent solving tasks in this category is distributed over a broad spectrum, from 6 seconds to 11.2 minutes, with an average of approximately 1.5 minutes per exercise. Again, the case of exercise 141 is notable, on which she spent 13.2 minutes and three attempts without solving it. The task asks respondents to find the generating expression and it involves multiplication of fractions, as can be seen in the following description.

> *Task 399*, high level. [0 attempts, resolution time 0. Input: a: ?; generator expression: $(ax+5)(-4)$; output: $-28x-20$]

This task can be difficult to solve by inspecting and comparing of expressions; however, finding the solution is easy by solving equation $(ax+5)(-4)=-28x-20$ for a. The correct answer is $a:7$. As for F's performance, no attempts or resolution time were recorded. However, he did perform the task and his work was probably entirely done with paper and pencil algebra.

> *Task 141* from "Fractions 2". [Inputs: a: 3; b: $\frac{8}{7}$; generator expression:?; output: 24. First attempt $a \times b$; second attempt: $a \times b + 1$; third attempt: $a \times b + 0$]

It is notable that L does not establish the equivalence between performing the given operations in expressions $a \times b$ and $a \times b + 0$, and tests with both.

In the Algebraic Expressions category, a significant number of exercises fail to solve them correctly and she does not try some of them. The wrong exercises are: 235, 232, 230, 220, 218, 206, 197, 198, 199, 190 (16% of the category); in general she makes two attempts. It is important to mention that L worked on some of the exercises twice, on consecutive days.

For example, she tried exercise 224 four times (this was the exercise with the highest number of attempts), on 2 days. Overall, she spent 5.5 minutes on the exercise.

> *Task 224* from "Expressions 2". [Inputs: a: $60x+403$; b: $60x-403$; generator expression: ?; output: $60x+60x$. First attempt: $(a+403)+(b-403)$; second attempt: $a \times b$; third attempt: $(a-b) \times (a+b)$; fourth attempt (correct): $a+b$]

L's response attempts are interesting because they account for different ways of testing additive and multiplicative relationships between constant and literal parts of algebraic expressions which, to our mind, account for a perception of the structure of the given expressions, but also of relevant operations for transforming them and reaching the desired result. L spent from 9 seconds up to 11.5 minutes on the exercises in this category, with an average of 2 minutes per exercise. Most of the exercises were solved within one or two attempts.

With regard to the Exponents and Roots category, L performs the tasks in times ranging from 5 seconds to 2.5 minutes, although she solved most of them in less than a minute. She devoted one to three attempts to the exercises. However, again there are a number of tasks that are not solved correctly, even after two or three attempts, such as 252, 255, 260, 261, 266, 267, 273 (20% of total tasks).

For example, task 273 from "Division", on which she made three attempts and spent 72 seconds but failed to solve it correctly.

Task 273 from "Division". [Input: $a : -2$; generator expression: a^{-2}; output:? First attempt: $\frac{1}{-a^{-2}}$; second attempt: $\frac{1}{-2^{-2}}$; third attempt: $\frac{1}{2^{-2}}$]

Based on L's attempts, we surmise that her strategy was based on inspection and testing of different variants of the use of negative exponents. However, in all her attempts there is a "double presence" of the effect of negative exponents on the structure of the expressions, since she not only writes the exponent base as the denominator of a unit fraction, but she also maintains the negative sign of the exponent "–2".

L successfully solves most tasks in the Operations Hierarchy category; she does so in one attempt and in general devotes 30 to 100 seconds on each one. In some cases the answer is incorrect, although she does not necessarily spend more time on those incorrect items. L incorrectly solves tasks 278 and 283, devoting two attempts and 80 seconds and one attempt and 3 minutes, respectively, on them.

As for Algebra-Multiplication, L solved almost all the exercises correctly (37 of 40 tasks) and does so in a single attempt, with times mostly ranging from 6 seconds to 2 minutes. L fails to properly solve tasks 311, 317, and 318 in the latter category. For example, task 311 "Multiplication", on which she made three attempts and spent 3.7 minutes, is described below.

Task 311 from "Multiplication". [Input: $a : ?$; generator expression: $-ax$; output: $-x^2 + 4x$. First attempt: $-4 + 4$; second attempt: $-x + 4$; third attempt: $-x^2 + 4$]

L's attempts indicate a procedure of inspection and comparison of the given expressions, which nonetheless was unsuccessful. An alternative to solving the task could have been to propose the equation $-ax = -x^2 + 4x$, factor the output in such a way as to obtain $-ax = -(x-4)x$, and conclude that $a = x - 4$. The latter is a strategy based on the use of syntax and that leads to obtaining the answer directly.

L successfully solves virtually all tasks in Algebra Notable Products and Factoring in times ranging from 12 seconds to 3 minutes, and mostly in one or two attempts. As already mentioned, the resolution of these tasks involves the direct application of rules or known algebraic identities, or the identification of notable products in simple expressions. She only failed to solve task 371 correctly, to which she dedicated two attempts over a span of 6 minutes. L's performance in these cases enables us to identify operational fluency of basic syntax, both numerical and algebraic (multiplications, manipulation of notable products, and factoring of second degree polynomials).

Like most participants, L also found it difficult to solve the Challenges tasks. In general, the times she devoted to tasks in this category were diverse: from 12 seconds to about an hour (as in the case of task 412, which she solved correctly after two attempts). However, in most cases L made only one attempt, which denotes great attention given to solving tasks in this category. Of a total of 46 tasks, she solved eight (approximately 17%) incorrectly and did not attempt another (415 and 441; both of which are find input type, high level). In particular, her performance in the following cases should be highlighted:

Task 385, high level. [1 attempt, resolution time 3.5 minutes. Input: a: ?; generator expression: $a - (2x + 3)$; output: $3x - 3$]

As mentioned above, this is a task with visual salient elements. Another strategy would make it possible to solve the equation $a - (2x + 3) = 3x - 3$, which is trivial considering the mathematical knowledge that the study subjects have at their disposal. The resolution of this task is interesting because, in contrast to F's three failed attempts, L solves it in a single attempt for which she took more than twice as long as F took in his attempts. Based on the operational characteristics of L's actions, we can assume that for the resolution of this task she resorted solving the equation perhaps using pencil and paper.

Task 398, high level. [2 attempts, both incorrect, resolution time 3.75 minutes. Input: a: ?; generator expression: $16a^2 + 24$; output: $4(x^2 + 1)$. First try $a = 4$; second try $a = 8$]

As mentioned above, this is a highly difficult task to be solved by means of ASS capabilities. But L's attempts are indicative of a strategy of inspection and comparison of expressions, albeit restricted to numerical parts.

> *Task 402*, high level. [3 attempts, resolution time 5.3 minutes. Input: *a*: ?; generator expression: $-4(9x - a)$; output: $-36x + 24$. First attempt: $a = -24$; second attempt: $a = 24$; third (correct) attempt: $a = 6$]

Based on L's attempts, we can say that her strategy was based on inspection and comparison of the output and generator expressions. However, it is notable that she did not consider the numerical factor that is explicit in the generating expression, which has clear visual salience. Since she did not make the most of that visual salience trait, she proceeded to test several options until he managed to establish the correct one.

> *Task 412*, high level. [2 attempts, both failed, resolution time 58.3 minutes. Input: *a*: ?; generator expression: $(9x + a)(9x + a + 1)$; output: $81x^2 + 171x + 90$. First attempt: $a = 80$; second attempt: $a = 90$]

It is to this task that L devotes the most time. This is a task with visual salience elements, since if one notices that it involves the input and its consecutive (*a* and $a + 1$), it can be compared with the factoring of the term independent of the output ($90 = 9 \times 10$) and thus the necessary value of $a = 9$ can be obtained. Although L does try to compare the given expressions she fails to coordinate the relationship of consecutive expressions with the factoring of the output's constant term.

AAS, MEx Tasks, and L's Attempts

While she takes longer and makes more attempts, L's actions reveal skill in calculating and using numerical properties, as well as in application of manipulative algebra techniques. These numerical and algebraic manipulative competencies are mainly deployed in the responses to tasks in categories 1-10, 1-100, and 1-1000, Fractions, Hierarchy of Operations and Algebra-Multiplication, Notable Products and Factoring.

Although in a significant number of tasks (e.g., tasks 385, 402, and 412, presented above) L does not take advantage of the visual salience traits provided by the expressions involved, she does make widespread use of strategies based on comparing the structures of the expressions involved, which implies capabilities to perceive and analyze the structure of expressions. The foregoing is observed above all in her resolution of the tasks contained in the Algebraic Expressions and Challenges categories.

However, this analysis and comparison of the structures of the expressions involved were not sufficient for L to correctly solve the type II tasks with visual salience (find the input) and high-level type III tasks (find the generating expression). Apparently in those tasks L failed to coordinate the comparison and inspection of the structure of expressions with logical thinking operations and reversible processes to solve them. Nonetheless, for some of these tasks she successfully resorted to her expression operation competences to meet the demands raised and be in a position to solve them, after several attempts and spending a good amount of time on them, as shown above.

Algebra Structure Sense in MEx: Notes on a Semiotic Interpretation

Below, we present an account of some theoretical tools of analysis and semiotic notions that allow us to formulate an (initial) interpretation of the task network provided by MEx as a *textual space* designed to promote signification processes of fundamental structural algebraic notions.

From the semiotic point of view (Filloy et al., 2008; Puig, 2008; Rojano et al., 2014), algebraic expressions, in particular, and mathematical texts, in general, do not exist in isolation and independent of each other. In fact, the reading and transformation of any mathematical text occur in relation to other texts, both mathematical and non-mathematical. When users of our study read algebraic expressions and solve the tasks proposed in MEx, they begin a journey through a network of relationships between algebraic expressions and their possible transformations. The act of reading an algebraic expression – or a mathematical text, in general – consists in the production of sense for the text by the reader; that is, tracking, analyzing, and continuing the relations between that text and other texts.

Working with MEx gives users the possibility to immerse themselves in this network of intertextual relationships derived from their own specific mathematical, linguistic, cultural, and social experiences. And it is precisely through sense production of the algebraic expressions and the tasks proposed in MEx that, in turn, we construct what we call the *meaning* of these mathematical texts. From our perspective, *meaning* "becomes something that exists between one text and all of the other texts to which it refers and with which it is related" (Rojano et al., 2014, p. 390). That is, we understand the construction of meaning (or *signification processes*) of mathematical texts in relation to their intertextuality as well.

MEx provides users with a series of heterogenic mathematical texts, made up of different strata of competences in the use of algebra language, and which present tasks involving recognition of the ASS and sense production for its solution. From the semiotic point of view, some features of MEx make it possible to surmise that it provides a textual space (Puig, 2008; Rojano et al., 2014) that enables the reading and transformation of the proposed algebraic tasks. Hence, on the one hand, MEx imposes semantic restrictions on users for reading and transforming tasks, while, on the other, it allows users to produce personal intertexts – unique and unrepeatable – through which they produce algebraic senses in their personal selections of tasks of the MEx they choose to solve.

From the semiotic point of view, the set of tasks presented to users in MEx is an "open" series of algebraic texts, and they build their journey through the series by their selection of the tasks they solve in each category. It is noteworthy to point out that from this standpoint, the set of such journeys promoted by MEx fosters signification (or resignification) processes of structural features of algebraic texts as well as fundamental algebraic notions such as equivalence, substitution, and equation.

This way of understanding users' work with the MEx, together with the actions analyzed in this chapter (Francisco and Lulú), allows us to propose a proper revision of the notion of ASS.

Revisiting the Notion of Algebra Structure Sense

In Chapter I, Rojano (this volume) reviews different perspectives related to structure sense in algebra and combines them into a broad notion, referred to as "[...] the ability to be aware of structural properties of an algebraic object", and which is taken as a starting point for development of the theme throughout several chapters of this book. In addition to encompassing Hoch's conceptual definition and the operational definitions of Hoch and Dreyfus, as well as the ad-hoc definition for the abstract algebra of Novotna and Hoch, this notion also includes the set of capabilities that subjects mobilize when addressing non-routine algebraic tasks, such as those identified in Francisco's and Lulú's responses in their work with MEx.

The types of attempts made by F and L when solving some of those tasks show procedures that differ from the direct application of rules or techniques. Such is the case of F's attempts in type II and III tasks, which indicate indirect procedures rather than direct application of a technique. In such indirect procedures, different types of capabilities are brought into play, such as in the case of type III tasks, which by their very design require a comparison of the structures of the algebraic expressions involved. This comparison brings into play the ability to analyze the terms and operations of the given expressions (in this case, the inputs and the output) based on the characteristics of the terms and operations of the expression that is sought to be built (the generating expression). A simple example of the above is task 199, intermediate complexity, in which the inputs a: $5x$ and b: $6x$, and the output are given $-x$; to construct the generating expression one has to revise the degree of the terms of the inputs and compare it with the degree of the output expression and imagine the operation that, applied to the former, would result in the latter. By the way, in his first attempt at solving this task, F proposes $b - a$ (incorrect) as a generative expression and then corrects it on his second $a - b$ (correct) attempt. In high-complexity type III tasks, comparative analysis can be very difficult unless there are elements of visual salience in the given expressions that help to anticipate the characteristics of the generating expression.

In L's case, the number of attempts, time invested, and her attempts themselves at high-level tasks of algebraic content reveal strategies based on the comparison of the expressions involved, even though she is not always able to arrive at a correct answer. These structure comparison procedures are an expression of ASS capabilities, in the broad sense (already mentioned) of the notion, in which the "ability to perceive" is replaced by the "ability to notice" the structure. In this regard, it should be noted that even if she does not have daily contact with symbolic algebra in her teaching activity as a primary school teacher the history of L's fluency in direct application of transformative algebra techniques and their interaction with a set of MEx tasks contributed to her mobilization of such capabilities.

As such, it can be stated that the algebraic work – such as that required by some of the previously discussed MEx tasks – gives rise to manifestations of structure sense that do not necessarily come from an immediate visual perception, given that such work requires comparative analysis of the structural features of expressions, as well as elements of logical reasoning or reversible thinking. In other words, the concepts explained in this chapter show aspects of transformative algebra that go beyond the limits of traditional symbolic exercises, and which also show the influence of this transformative activity in other algebraic activities, such as generative and global/meta-level activities, characterized by C. Kieran (2014). In turn, it shows that combinations of these activities are conducive to the emergence of ASS capabilities that have not yet been well explored.

With regard to the contribution of technology, it is important to refer to the role played by technology in the emergence or development of such capabilities, both in the design by levels of complexity of individual tasks and in the design by "leveling up" of the task network. These design elements, which allow for users to make their way through the tasks depending on each one's mathematical experience, can be interpreted as characteristics of a textual space, which is underpinned by the intention to promote certain readings of the texts that conform it. According to the elements of semiotic theory presented above, reading/transformation acts based on interacting with a textual space immerse the reader into an intertextual network composed of their knowledge and linguistic experience. This intertextuality, in turn, gives rise to signification processes or resignification of the textual space itself and related texts.

From this perspective, MEx task network and tasks form a textual space designed to promote structural reading/transformation acts of expressions and relations between algebraic expressions. The theoretical hypothesis in this case is then that through these reading/transformation acts signification or resignification processes related to structural features of algebraic texts are fostered, by immersing the user (reader) into a network of intertexts formed by his/her mathematical (algebraic) experience. The description of principles upon which the design of MEx components is based, as well as the indications of the presence of ASS in the cases of F and L when working in this technological environment contribute to strengthening the above hypothesis. Clearly, strategies involving inspection and comparison of the structures of algebraic expressions account for the construction of meanings and resignifications of texts that are the expressions themselves and their interrelations.

To date the arguments and discussion aimed at deepening the notion of ASS have referred to the particular case of the design of and experience with MEx. However, for that purpose it will be important to consider other developments as well, such as several that are described and analyzed in Chapter 4, the designs of which suggest conceptions on the development of algebraic capabilities akin to those discussed here.

Notes

1 Visual salience features are elements (markers) that are visually effective for gaining awareness of the structure of an expression.
2 A three-year Frontiers of Science project funded by the National Council of Science and Technology (Conacyt) in Mexico, Ref. 2016-01-2347.

References

Filloy, E., Rojano, T., & Puig, L. (2008). Educational algebra. *A theoretical and empirical approach*. New York: Springer. https://doi.org/10.1007/978-0-387-71254-3

Filloy, E., Rojano, T., & Solares, A. (2003). Two meanings of the equal sign and senses of substitution and comparison methods. In N. A. Pateman, B. Dogherty, & J. Zilliox (Eds.), *Proceedings of the Twenty-Seventh Annual Conference of the International Group for the Psychology of Mathematics Education* (Vol. 4, pp. 223–229). Honolulu, Hawai'i, USA: CRDG, College of Education, University of Hawai'i.

Filloy, E., Rojano, T., & Solares, A. (2010). Problems dealing with unknown quantities and two different levels of representing unknowns. *Journal for Research in Mathematics Education*, *41*(1), 52–80. Retrieved July 14, 2021, from http://www.jstor.org/stable/40539364

Hoch, M. (2007). Structure sense in high school algebra. *Unpublished doctoral dissertation*. Israel: Tel Aviv University.

Kieran, C. (2014). Algebra teaching and learning. In S. Lerman (Ed.), *Encyclopedia of mathematics education* (pp. 27–32). Dordrecht: Springer. https://doi.org/10.1007/978-94-007-4978-8_6

Peirce, C. S. (1982). On the nature of signs. In C. J. W. Kloesel (Ed.), *Writings of Charles S. Peirce: A chronological edition* (Vol. 3, pp. 66–68). Bloomington, IN: Indiana University Press.

Puig, L. (2008). Sentido y elaboración del componente de competencia de los modelos teóricos locales en la investigación de la enseñanza y aprendizaje de contenidos matemáticos específicos. *PNA*, 2(3), 87–107. doi: https://doi.org/10.30827/pna.v2i3.6199.

Rojano, T., Filloy, E., & Puig, L. (2014). Intertextuality and sense production in the learning of algebraic methods. *Educational Studies in Mathematics*, 87(3), 389–407. doi: https://doi.org/10.1007/s10649-014-9561-3.

6 Students' Justification Strategies on the Correctness and Equivalence of Computer-Based Algebraic Expressions

Eirini Geraniou and Manolis Mavrikis

Introduction

As many authors (e.g., Arcavi et al., 2017; Mason, 1996; Radford, 2011) have argued arithmetic is the precursor of and prerequisite for algebra and even though algebra is considered "generalised arithmetic", arithmetic and algebra have different foci. Arithmetic encourages students to find numerical answers, whereas algebra encourages students to identify and express mathematical structures. While algebra is a system that supports structural sense and expresses generalisation, its teaching often prioritises the acquisition of its transformation rules, rather than structural thinking and generalisation itself. There have, of course, been numerous attempts to address this challenge, and more generally, to come to grips with the difficulties that students face in their transition from arithmetic to algebraic thinking. For example, activities based around generalising patterns of various descriptions have been widely considered as a potentially powerful way to help students learn how to "see the structure" and generalise and are used as a common route for introducing algebra (e.g., Küchemann, 2010; Mason, 1996; Radford, 2010; Radford et al., 2007). However, the use of these activities needs care, as they can easily encourage trial-and-error techniques and a focus on the term-to-term (or in other words additive) rule, which do not necessarily lead to mathematically valid generalisation strategies that promote the acquisition of structural sense for the learner (Dörfler, 2008; Hart, 1981; Küchemann, 2010; Küchemann & Hoyles, 2009; Radford, 2011). The challenge, therefore, is to introduce patterns and generalisation in ways that promote algebraic thinking, that is, to identify and express structural commonalities and relationships (Geraniou et al., 2009; Noss et al., 2009).

Our aim is to investigate the impact of collaborative computer-based tasks involving figural patterns to students' justification strategies for discussing the equivalence and correctness of algebraic expressions and subsequently students' development of algebraic generalisation and structural sense in Algebra. In the analysis of our empirical data involving such tasks, we will be discussing the interconnectedness of figural

DOI: 10.4324/9781003197867-6

and numerical patterns for the development of algebraic structural sense and generalisation.

Different authors have defined mathematical generalisation in various ways: "as a mathematical rule about relationships or properties" (Ellis, 2007, p. 196); as the process of extending one's scope of reasoning beyond the case or cases considered (Harel & Tall, 1991); or as communicating at a level "where the focus is no longer on the cases or situations themselves, but rather on the patterns, procedures, structures, and the relationships across and among them" (Kaput, 1999, p. 137). More recently, Radford (2010) argues, "generalizing a pattern algebraically rests on the capability of grasping a commonality noticed on some elements of a sequence S, being aware that this commonality applies to all the terms of S and being able to use it to provide a direct expression of whatever term of S" (Radford, 2010, p. 42). Ellis (2011) argues similarly, but adds a reference to the sociomathematical contexts in which students are engaged.

In this chapter, we refer to a restricted class of generalisation: namely, the process of noticing the structure of a figural pattern, identifying what is repeated, and expressing the rule that corresponds to this structure symbolically. Our interest also encompasses the role of justification of generality in a carefully designed collaborative social context, as a step towards the adoption of algebraic ways of thinking.

While it is relatively straightforward for students to illustrate some elements of generality using numbers and gestures, expressing generality in words or in algebraic form has proved more problematic (Arcavi, Drijvers, & Stacey, 2017; Filloy and Rojano, 1989; Noss, Healy, & Hoyles, 1997; Noss et al., 2009; Radford, 2010, 2011;). Frequently, students view the algebraic expression as disconnected from the structure of the problem, to be added as an optional and seemingly arbitrary endpoint.

In this chapter, we present a research study, which aimed at bridging the gap between identifying and expressing pattern, by encouraging students to identify the structure of the pattern *through its construction* and providing them with the necessary tools to express generality. Thus, the underlying theory that guided our activity design is constructionism (see Harel & Papert, 1991) and the claim is that this experience will support the expression of generality by focusing on the structural sense in Algebra, as well as shape how the generality is expressed. We used the microworld, eXpresser, a toolkit for working on the construction of tiling patterns (more details given later), and activity sequences that ended in a reflective and a collaborative phase aiming to provide students with a rationale and opportunity to justify their pattern constructions to each other. We focus on this final phase of the activity sequences. We present data from several studies in three English schools, illustrating how 11- to 14-year old students who had engaged with eXpresser were able to reflect upon their own and their peers' solution strategies and employ a range of strategies to justify the correctness and – where appropriate – the equivalence of their computer-based algebraic rules. Our research questions were: What is the impact of computer-based collaborative tasks involving figural pattern generalisation on students' justification strategies for the equivalence and correctness of algebraic expressions? How do such tasks support the development of algebraic generalisation and structural sense in Algebra?

Theoretical Background

Perceiving and Expressing the Structure of Figural Patterns

The pattern-generalising process involves various steps that students typically follow in their efforts to reach meaningful generalisations. Dretske (1990) argues that the initial act of coming to see a pattern is of two types: sensory, which refers to individuals perceiving an object as a mere object-in-itself; and cognitive, when perception moves beyond sensory perception by recognising a fact or a property related to the object in question. We will argue that Dretske's distinction between these two modes might not be so clear-cut, depending crucially on the tools at hand, and the contexts in which they are used. A key issue, however, as Rivera (2010) points out, is seeing or recognizing a fact or a property in relation to an object, and doing so cognitively or theoretically rather than practically. In fact, the way in which the structure of a pattern is conceived will depend critically on the unit of repeat perceived, and – this is the crucial point – the tools available for repeating.

If and when students recognise what is repeated, they are often capable of expressing a general rule through the use of words like "always" or "every", but struggle to use letters and symbols (see Warren & Cooper, 2008). Additionally, Radford's (2009) discussion of "objectification" shows how students' inexperience with mathematical language prevents them from "translating" the structure of a pattern to an algebraic expression after noticing what is repeated). To address this, Rivera (2010) advocates an *"abductive-inductive action on objects"* that involves "employing different ways of counting and structuring discrete objects or parts in a pattern in an algebraically useful manner" (ibid., p. 4). The idea is that such an approach allows students to perceive the structure of a pattern that could lead to a meaningful, but also mathematically viable, pattern generalisation, which could form the basis for deriving a formula.

After this initial step of recognising the structure of the pattern, students are usually encouraged (e.g., in mathematics lessons by a teacher) to adopt a suitable symbol system to reason about and to express generalisations (Arcavi et al., 2017; Kieran, 1989), which is once again a far from trivial process. This action has been characterised as *"symbolic action"* by Rivera (2010) and involves translating the visually perceived structure of a pattern into the form of an expressed generalisation. This transition from arithmetic to algebra, or "didactical cut" as described by Filloy and Rojano (1989) involves a substantive shift from operating with numbers and operating with the unknown (see also Filloy et al., 2010). To effect this transition, Filloy, Rojano, and Puig (2008) argue that a teacher's intervention is crucial.

The pedagogical strategy used by Filloy, Rojano, and Puig (2008) involved two concrete models: the *balance scales* model, in which the equation was represented as a balance between two weights in two pans, and the *geometric model*, in which the letters and algebraic expressions were represented as lengths and areas of rectangles. The lesson we take from this work is that while a modelling approach can support the development of algebraic thinking, it can also become a hindrance in the absence of carefully designed activities to support pedagogically the transition to algebraic expression.

Algebraic expression, of course, does not have to be verbal or written. Noss and Hoyles (1996), for example, use the idea of *situated abstraction* to describe how students can express abstractions *within* a "concrete" symbolic medium, such as a computer program; Radford (2010) similarly invokes the notion of *semiotic means of objectification* to capture the means used by students to express a general rule, such as gestures, signs, etc. Designing semiotic systems is therefore a worthwhile challenge in the attempt to foster the expression of generality. However, the utility of such systems is sensitive to their structure: there are, for example, several studies (e.g., Lee, 1996; Stacey, 1989) that highlight students' tendency to focus on recursive rather than functional relationships, which can present barriers towards generalising "any" case. Furthermore, generalisation tasks that are presented as a sequence of consecutive terms often lead students to seek empirical generalisation rather than a structural one (Bills & Rowland, 1999). More recent research documents the different strategies students employ when constructing the algebraic rules that underpin patterns, which allow them to support the correctness of their general rules. For example, Rivera and Becker (2008) differentiate between constructive and deconstructive generalisation, depending on whether or not students perceive the figural pattern as having overlapping components, and Chua and Hoyles (2011) refer to *reconstructive* generalisation, where components of the pattern are rearranged to reveal the pattern structure.

Collaboration and Reflection

Alongside the affordances of novel representational systems for describing generality, algebraic thinking can be further supported by opportunities for reflection, to assist students in expressing structural relationships and distinguishing variants and invariants (see for example, Ellis, 2007). Encouraging students to reflect on their actions can promote their justification skills and support the development of their algebraic thinking.

Unsurprisingly, reflection can be strengthened by suitably designed collaboration tasks. As students explain their ideas and solutions to their peers, they are encouraged to resolve conflicts and develop a deeper understanding than those who do not (Cohen & Lotan, 1995; Leonard, 2001; Lou et al., 1996). Wood (1988) argues that discussion and interaction are critical to avoid "misconceptions": "A trouble shared, in mathematical discourse, may become a problem solved" (p. 210). When collaborating in small groups, students are more likely to ask questions, reflect on their own work and attempt to make sense of each other's work (Linchevski & Kutscher, 1998). In the context of generalisation, research suggests that working in small groups is advantageous for students' deeper understanding of generalisation, equivalence of rules and algebraic thinking (Ellis, 2011).

Leonard (2001) argues for the value of forming heterogeneous groups, especially for lower-attainment sixth grade mathematics students who were grouped with higher attainment students and achieved more although other disagree (see for example, Carter & Jones, 1993, as cited in Fuchs et al., 1998). Criteria such as these for grouping students are dependent upon the mathematical tasks they are asked to tackle and the learning objectives assigned by the teacher. Furthermore, students who work in small groups seem to learn more when the outcome depends upon all of the group members' efforts (Lou et al., 1996). A cooperative learning strategy encourages students to share arguments and consider different approaches that could even be shared with the rest of the class. Hopefully such an approach minimises the issue of one student overpowering the group. The question is how to *design* group work so it is most likely to lead to optimal results for learning (Abdu, Olsher & Yerushalmy, 2019; Healy et al., 1995).

Argumentation and Justification

Even though students are capable of generalising a pattern or a rule, few are able to explain why the rule is valid and justify their actions (Coe and Ruthven, 1994; Ellis, 2007; Healy and Hoyles, 2000; Küchemann and Hoyles, 2009). Many tend to rely on empirical examples to justify the truth of statements: it would hardly be surprising if a student who generalises based solely on specific cases were to use one or more examples as a form of justification.

Research suggests that a student who generalises by attending to the structure of a pattern and relating each algebraic expression to its corresponding part of the pattern-model-construction, has a better chance of justifying the generality of their expression and possibly producing a general argument to justify the equivalence of rules (Küchemann, 2010). Helping students develop their own algebraically powerful generalisations will likely aid in their abilities to provide symbolically-expressed justifications or in other words some form of proof (Ellis, 2007). Thus setting challenges for students to reflect, recognise, and justify general rules and actions to themselves as well as to others, is a strong candidate for a strategy to enhance students' generalisation capabilities. Moreover, it seems that focusing on justification activities may not only enhance students' expression of their existing generalisations, but also aid in the development of subsequent, more powerful generalisations (Ellis, 2007).

Mercer (1995; 2000, as cited in Swan, 2006) has shown that attention should also be directed at promoting exploratory talk (critical and constructive exchanges) among the participants of the group rather than disputational (disagreement and individualised decision-making) and cumulative talk (build positively on others' input). Working collaboratively (rather than competitively), students tend to be more committed to overcoming conflict in their efforts to master a task and coordinate different pints of view into new ones (Laborde, 1994). Furthermore, promoting a collaborative "knowledge building" culture as envisaged by Scardamalia and Bereiter (2006) during collaboration, can advance students' knowledge.

Collaborative Interaction in Relation to Computers

There is a long research tradition in collaborative learning with computers in mathematics education. Early work from Teasley and Roschelle (1993) and Healy et al. (1995) highlighted the importance of expressing ideas in words and establishing a common group goal. Similarly, a key research finding is that a characteristic of computational artefacts is that students' focus of attention can gradually change from being computer-oriented to being focused on the mathematical aspects of the task at hand (e.g., Kieran, 2001; Lavy & Leron, 2004).

Among other relevant findings, Bereiter and Scardamalia (2003) and Moss and Beatty (2006) discuss how students make progress not only in improving their own knowledge but also in developing "collective knowledge" by contributing to their peers' comments. The WebLabs project took this a step further through its "Webreports" system (Mor et al., 2005) that facilitated distance collaboration of students who constructed models. This study demonstrated the positive effects of sharing, commenting, making changes and allowing students to reflect on each other's artefacts both synchronously (face-to-face) and asynchronously (Mor et al., 2006). The importance of students' engaging with, or talking about, the product of their work and the opportunities for building on each other's ideas, learning to participate in a community of practice and benefit from the reflection that occurs from the interaction with others has also been widely identified (Vahey et al., 2000; Vahey et al., 2007).

Ellis's (2011) research on the role of collaboration in the development and refinement of students' mathematical generalisations highlighted how generalisation can be viewed as "a dynamic, socially situated process that can evolve through collaborative acts" (p. 308) where, in the classroom situation outcomes are influenced by the interconnected actions of students, teachers, problems, representations, and artefacts. In her research, students were stimulated publicly to generalise, share, build, and encourage justifications or clarifications in their attempts to understand a new mathematical domain. Moreover, her research revealed that an initial generalisation evolves through extensive inter-actions, phases of reflection, and takes many different forms until the final, stabilised version of generalisation, which cannot be claimed to have been developed in isolation. Rather, every student in a group is responsible for the final generalisation. This is described as collective generalising (ibid., 2011).

The eXpresser Microworld and Activities

The key idea of the eXpresser microworld is that students first *identify* the structure of a pattern of squares presented dynamically, next *construct* the pattern, and finally *express* a general rule for the number of tiles in a general pattern. Thus, there is a tight coupling between building the pattern, and being able to describe how it is built – between the "algebra" and the objects the algebra aims at expressing. The quotes round the word algebra signify, as we will see, that the language of expression is algebra, but not as we know it.

For our purposes here, we see students' work in eXpresser to solve a generalisation activity as involving two phases, the construction and the collaboration phases. In the construction phase, students go through the following actions: (1) visual perception of the model presented, (2) inductive action on the model, to realise what stays the same and what is repeated, (3) constructionist action, building the model (using one or more patterns), (4) expressive action, expressing the constructed model in the form of a general symbolic rule that colours the model. Then in the collaboration phase, students continue their interactions with: (5) reflec-tive action with students writing arguments for or against their models and particularly their derived rules (these arguments are written at the end of the construction phase and used during subsequent collaboration, (6) justification action, viewing other students' rules and validating/justi-fying correctness and equivalence, and (7) collaborative reflective action, involving students in groups to reach an agreed statement regarding the equivalence and correctness (or not) of their shared rules.

A typical activity in eXpresser will ask the student to reproduce what is presented as a dynamic model of a tiling pattern shown in a window that appears on the side of the activity screen. In eXpresser, an initial figure is presented dynamically in order to draw students' attention to the general problem, rather than the static and inevitably therefore specific problem that could otherwise be posed on paper. Figure 6.1a shows the "Train-Track" model: it is animated randomly from left to right[1] with the value of the model number changing accordingly.[2] Students are asked to construct the Train-Track model in eXpresser using patterns and combinations of patterns of their own choosing (examples are shown in Figure 6.1), depending on their perceptions of the Train-Track's structure. Students are encouraged to colour the patterns with different colours to represent the way they visualise the structure of the model. This happens by providing an expression that "tells the computer" the number of tiles in each coloured pattern. They then seek to derive a general rule for the total number of tiles needed for *any* Model Number. The generality of the rule can be tested by "animating" the figure to see if the model matches the activity model, is drawn correctly, and remains coloured for any model number the computer chooses randomly.

Since the activity model is presented dynamically, as an animated pattern, students are given the opportunity to perceive the model visually and construct it in any way they see it. They can therefore identify the structural relationships in the model in an abstract way that would materialise as they carry on constructing the model and make meaningful generalisations. Based on their visual perception of the activity-model, students are expected to derive a rule that expresses their method of construction. Students identify a common unit in the activity-model, build it, and repeat it to form a pattern. A number of patterns can make their model. In an effort to discourage students from thinking from term-to-term and therefore additively, eXpresser is designed to help students to express the relationship between the common unit and the number of its repetitions. This common unit is referred to as a "building block": the idea of using

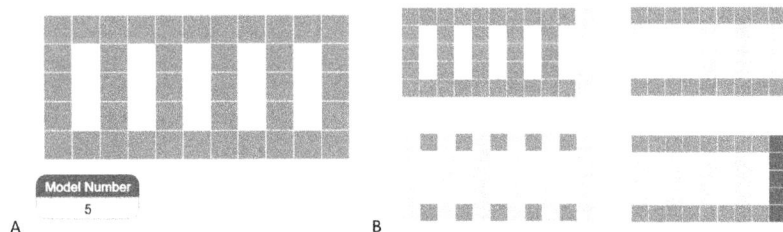

Figure 6.1 (a) The Train-Track activity and (b) six different students' perceptions of it.

different building blocks for different patterns comes from the prevailing view in recent studies involving figural pattern generalisation activities that students see a pattern in different ways depending on how they conceptualize such units (see the ZDM issue on pattern generalisation for references to most studies; Rivera & Becker, 2008). This process comprises the second step in the generalisation activity diagram. This common unit is the coefficient (constant) of students' variable in the general rule and as it will be shown later in this chapter, students employed the notion of the constant to justify the correctness of their rules.

The eXpresser capitalises on visual dynamic representations and feedback[3] and, in addition, on the simultaneous representation of a specific and a generalised model, called "My Model" and "General Model" (see Figure 6.2 presenting the model of a year 8 student, Alicia). The model is built by combining patterns and there is a close alignment of the symbolic expression, the Model Rule and the structure of the model. In the General Model, a value of the variable[4] ("Model Number" in this example) is chosen automatically at random (it is "6" in Figure 6.2. It will generally be different from that in the specific model ("3" in Figure 6.2).

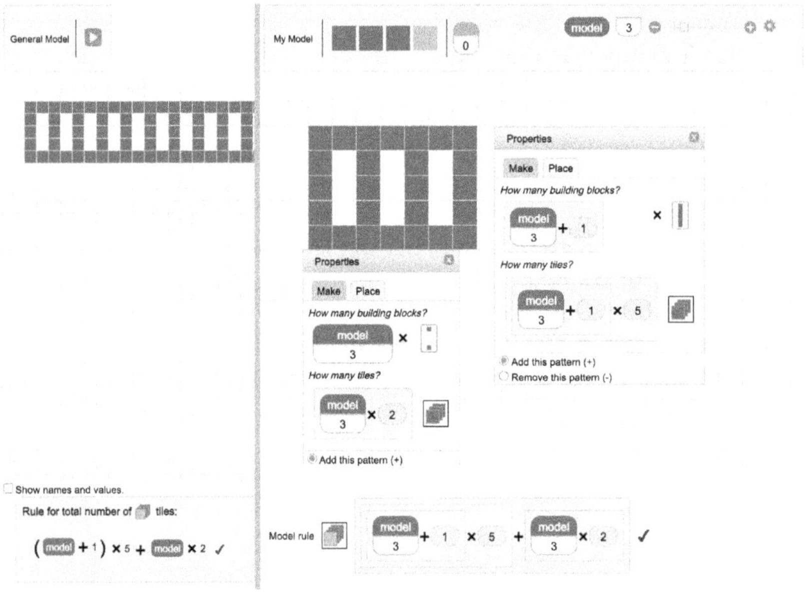

Figure 6.2 Alicia's Train-Track model showing her constructed model on the right, where she chose her variable "model" to be 3. On the left hand-side, there is the General Model window, where her constructed model is demonstrated for the value of 7 for the 'model' variable. Also, below Alicia's Model, one can see her derived general rule given in the eXpresser language. On the General Model Window, Alicia's Model rule is presented in an algebraic form.

So the General Model indicates to students whether their constructions are structurally correct for the different values of the variable(s) assigned to the various properties.

The system is designed to encourage students to express the structure of a pattern using the eXpresser representational form (which is the "Model Rule" in Figure 6.2). Students construct a model rule for the total number of tiles, and validation of its correctness is made evident by the system through colouring: patterns are *only* coloured if the rule for the number of tiles required is correct. Motivation for generality is thus provided by the main goal of the activity, that is, to produce a model that will animate correctly by colouring the exact number of tiles required, in combination with a pedagogical strategy that challenges students to construct models that are impervious to changes in the values of the variables.

Designing for Collaboration in eXpresser

Students' notions of generality, even when constructed in an environment designed to scaffold mathematical generality, might not match what it is required for a meaningful generalisation. We therefore designed individual reflective activities, in which students were required to think of and write down their arguments for their rules' correctness. They were prompted to think carefully how best to explain and justify this in written form, to encourage them to come up with good arguments and express them explicitly. This activity would also prepare them for the second part of the collaborative activity where they discuss the equivalence of different rules.

As illustrated in the final stage of the MiGen Activity Diagram (Figure 6.1), students were grouped in pairs by their teacher with the advice of the Grouping Tool (for a description see later) based on their different, yet equivalent, derived rules. During their collaboration, processes of ascertaining and persuading each other of their rules' correctness and equivalence were revealed. In fact, there was a process of ascertaining when a student was going through the reflection phase, which was followed by a process of persuading when they were paired during the collaboration phase. Also, collaboration encouraged the development of reasoning skills since students were encouraged to explain why their generalisations make sense.

Having reviewed the literature, we based our approach on the assumption that collaborating *about* and *with* something concrete (in this case, virtual) is more likely to lead to effective learning than without. The activities were designed to prompt students to revisit their derived rules and through discussion with their peers reflect on their transition from simple to more complex expressions and then to algebraically accepted ones.

We conjecture that focusing on figural pattern activities with tuned tools and carefully structuring and providing a context for students' discussions on the correctness and equivalence of models and related algebraic expressions could prove to be a powerful approach to fostering algebraic thinking. In the context of this carefully designed exploratory learning environment eXpresser, students are given the means to relate the symbolic representation to the relevant parts of the pattern, give meaning to symbols, and form justifications in an algebraic manner.

Interactions in eXpresser

To give the reader some understanding of students' interactions with the eXpresser microworld, we will briefly outline Alicia's interactions with the Train-Track activity before we move on to consider the research methodology employed mainly for the collaborative activities.

Alicia was presented with the animated Train-Track model and was asked to construct it and find the general rule. She placed five red (dark grey in Figure 6.2) tiles to form a column. She wanted to repeat this column a number of times towards the right direction, but always leaving a gap in between each two consecutive columns. She therefore characterised this column in eXpresser as a building block. She decided to repeat this building block three times to make her blue pattern. She then made a building block of two green (light grey in Figure 6.2) tiles placed vertically, but with a gap of three tiles in between, and repeated it three times. She noticed that the end of her model did not match the activity model as there was a column missing. She then derived a relationship between her two patterns that for **every number of green building blocks, she always needs an extra red building block**. Using the eXpresser's language she was able to express this relationship. For any number of green building blocks, for example [model:3], she needs [model:3] + 1 red building blocks. To colour each of her patterns, she had to multiply the number of repeats for the building block with the number of tiles in one building block. So, for the green pattern, she built the expression [model:3] × 2 and for the red pattern, {[model:3]+1} × 5. Since the activity was to find the rule that gives the number of tiles in the model for any model number, she added the two expressions that coloured her two patterns and got her general rule: {{[model:3]+1} × 5} + {[model:3] × 2}.

We now return to the study itself, and focus on the methodological approach regarding the reflective and collaborative aspects of our study.

Methodology

The purpose of our research study was to collect data that illustrate how students who had engaged with eXpresser were able to reflect upon their own and their peers' solution strategies and employ a range of strategies

Table 6.1 Sample

Schools in London	Year group	Number of students in class	Number of students interviewed during collaboration
A	Year?	20	6
B	Year?	24	6
B	Year 8	22	6
C	Year 8	16	16
C	Year 9	28	14
TOTAL			48

to justify the correctness and – where appropriate – the equivalence of their computer-based algebraic rules. Such data would allow us to discuss the impact of computer-based collaborative tasks involving figural pattern generalisation on students' justification strategies for the equivalence and correctness of algebraic expressions. Subsequently, we would make inferences regarding the degree to which such tasks would support the development of algebraic generalisation and structural sense in algebra.

The data presented in this chapter are derived from 48 mixed ability year 7, 8, and 9 students (aged 11–14 years) from three different schools in England. Table 6.1 presents the number of students interviewed from each school.

The following sections describe the sequence of activities students undertook, how the Grouping Tool, another tool of the MiGen system, paired students, and the data analysis process that was followed to reach our results.

Students' Activities in eXpresser

Students were familiarised with the eXpresser in two lessons through a number of introductory activities and practice activities, asking students to construct figural models. Afterwards, they were given the Train-Track activity and were asked to build the model on their own. The activity text comes with suggested goals as follows: "Construct the Train-Track model. Use more than one pattern to make the model. Use different colours for each pattern to show to other people how you made your model. Find a rule for the number of tiles for any Model Number. Use pattern(s) to construct the model; make sure "My Model" is always coloured; check that the "General Model" animates without messing up; make sure that the "General Model" is coloured always". Students constructed the model in many different ways, some of which are presented in Figure 6.1.

After constructing their model, students were asked to reflect on it (reflective action) by answering the following questions: *(1) Use your*

*model to find the number of tiles for Model Numbers: 6, 12, 1, and 100.
(2) Is your rule correct or not? In the next activity, you will discuss with
another student. Make some notes here to explain why your rule is correct or not to prepare for this group activity.* Based on having different
student models and rules in any pair, students were paired to discuss the
correctness and equivalence of their rules as mentioned earlier.

Students' Collaboration

Student pairs were formed by exploiting information provided by the
Grouping Tool,[5] which was designed to support the teacher in deciding
upon the best possible pairs of students in terms of potentially worthwhile discussions on the equivalence of eXpresser rules. The grouping
tool retrieves all students' models from the database and analyses them
on the basis of three criteria: (1) the similarity of the building blocks,
(2) the values of the right and downward displacement of the building
blocks, and (3) the similarity of expressions that relate properties of the
models (e.g., that a building block is repeated twice as another). Based
on its analysis, it suggests to teachers' possible groupings based on the
dissimilarity of students' models. The final decision for the best pairings
of students of course lies with the teacher, so the tool is designed to allow
the teacher to change the suggested pairs, as they deem appropriate.

Having been assigned to their pairs the students were asked the
following two questions: *(1) Convince each other that your rules are
correct, (2) Can you explain why the rules look different but are equivalent? Discuss and write down your explanations.* Students were asked
to write their final arguments on paper to share with the rest of the class,
to stimulate discussion orchestrated by the teacher.

Data Collection and Analysis

For each school study, based on the grouping tool's and the teacher's
suggestions, a number of pairs of students were chosen to be interviewed outside the classroom during their collaboration (in some cases
we interviewed all students, whereas in other cases we interviewed at
least half a class). The rest of the students worked in pairs in the classroom. This decision served our research purposes as it allowed us to
better insight into their ways of thinking in a quiet space (outside a
noisy classroom) and where their discussions could be recorded.[6] In
the interviews, the two students' models together with their rules were
opened in eXpresser on the same machine so the pair could explore and
interact with the models if they wished. A snapshot of each student's
final model and rule in eXpresser was also presented colour-printed
on paper in front of the students. Students' interactions were vid-
eo-recorded and their verbal discussions recorded, transcribed, and

analysed. The students were encouraged to write some arguments to prepare for their discussions, as mentioned earlier, and this written work provided more data as to how they expressed their arguments in written as well as verbal form.

When analysing the 24 transcripts, our focus was on the students' strategies in supporting their arguments to their peer. By the time the study was undertaken, students in years 8 and 9 had been introduced to generalisation activities in their normal classes whereas this was not the case for the younger year 7 students. But, as we understood from the teachers, no student had participated in any discussions on the topic of equivalence of algebraic expressions. Our interest, therefore, extended to seeing how students would reflect upon their general rules prior to the collaborative activity as well as during the discussions with their peers and assess their understanding of a general rule.

Throughout their interactions with eXpresser, support was provided through the intelligent support of the system, the teacher and the researchers. During collaboration, the researcher's role was to initiate the discussions by introducing the activity and intervened only to clarify issues that were raised by the students by asking them to provide more details for their explanations. The researcher encouraged students to ask their fellow student questions if they were unsure of the other's rule or their arguments. They reminded occasionally students of their collaborative activity and prompted them to clarify their arguments or ideas to their peer (especially if they seem confused and didn't take the initiative of asking if they were unsure).

The students' discussions, orchestrated by a researcher as described above, were audio-recorded, transcribed, and analysed qualitatively following the *open coding, axial coding*, and *selective coding* analytical processes as described by Strauss and Corbin (1998). To facilitate this process, the Transana[7] software was used, which allowed us to annotate all the transcripts and create codes that later formed categories and themes. During the open coding process, the transcripts were conceptualised line-by-line to create codes that were constantly compared, renamed, modified, or even merged into new concepts so as to reach a saturated list of codes. After constant comparison of the data and several modifications of the derived concepts, and an intent to sharpen the emerging theory, a number of categories that described the justification strategies students were using to argue about their rules' correctness and equivalence were generated. Once a first set of categories had been established, the raw data were revisited to assess their validity and evaluate whether all the justification strategies used by students were adequately captured. When this was not the case, a new category was incorporated and validated against the data. This iterative cycle was repeated a number of times until the set of justification strategies was finalised.

After the categories and subcategories were finalised and we reached what Straus and Corbin refer to as "theoretical saturation" (Strauss & Corbin, 1998), students' different justification processes were grouped under two themes: (a) Justification for Correctness and (b) Justification for Equivalence. It was at this point that we integrated all major categories and formed a coherent theoretical scheme (*coding scheme*) to present our conceptualisation of students' approach towards justifying their generalisations. In what follows we outline the classification of strategies based on the coding scheme in its final state and show how it relates to the theoretical framework we outlined at the outset.

Justification Strategies

Justification for Correctness

During the first part of the collaborative activity, students in their pairs were asked to convince each other that their rules are correct. In their efforts to give reasons for their rules being correct and therefore convince their pair, all students referred to the arguments they had written during the reflective action from the construction phase as preparation for their discussions and used them to start off their discussions.

During their collaborations, students used a number of different strategies as identified in the data. Note that 18 out of 48 students used more than one strategy to justify the rules' correctness. Some students used their models to match each part of the rule to the corresponding coloured pattern of their model. Such an approach was characterised as *constructive justification* (number of responses: 26). For example, a Year 7 student, Janet, said for her model (see Figure 6.3):

> Janet: that's the number of red. But there's 3 tiles in each of the blue building blocks, and the one plus 6 is because the 6 is the number of these and there's always one more here, there's one more of blue building blocks than the red because of the green one on the end and then the 2 green ones.

Other students used their models to explain the coefficients in their rule based on the number of tiles in each repeated building block. Such an approach was characterised as *structural justification* (number of responses: 29). For example, a Year 7 student, Patricia, said:

> *Patricia:* we have the blue bits, which is 2 of them, each are blue, and then we have the red block. And the other 2 tiles. So that makes 4 blue [...] then add it together with red. Add these two together.
> [That's the same with Emily's because she is talking about the C].

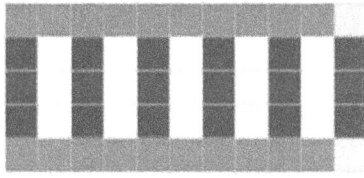

Janet's model

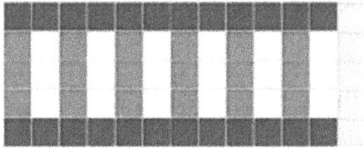

Patricia's model

Emily's model

Figure 6.3 Three example models.

In both Patricia and Janet's strategy, we see traces of the construction each had pursued, and in both cases, the justification clearly took for granted that the audience (i.e., the other student) had participated in a similar activity. Indeed, in these cases, it is not difficult to share the situatedness of each abstraction, in which the constructive aspects of the spoken generalisation are expressed in the medium of the activity (e.g., "add it together with red"). These students directly related either the different patterns in their models to the different terms in their rules, or the number of tiles in their constructed building blocks to the coefficients in their rules.

As we might expect, students often used examples as a "proof" for their conjectures. Such an approach was characterised as *empirical justification* (number of responses: *10*) and was chosen by a few students as a strategy that allowed them to test their rules with the help of the eXpresser's feedback, that is, if their rule was correct, choosing a different value for the independent variable would maintain the model's colouring. For example, a Year 7 student, Emily, said:

Emily: We answered correct when we typed in the model number.

Emily tried a number of different values for the model number and each time her model remained coloured and did not "mess-up".[8] The idea of the computer "not being wrong" seemed to play a crucial factor in her judgement as in addition to her tendency to focus on a few cases to generalise for *any* case, she cleverly relied on eXpresser's immediate feedback. During her collaboration with her pair, Patricia, Emily was challenged to use a different strategy later in their discussions that of constructive justification, as is evident from the results presented in Table 6.2, described below. Similarly, Susan and Tod, the only Year 8 students who used the empirical justification, only used this strategy initially, and later relied on a different one as their collaboration in their pairs progressed. In fact, all the Year 9 students used the empirical justification strategy as an additional one to the ones they used initially (which were mainly the constructive or the structural ones), to support further their arguments with the support from the eXpresser's immediate feedback.

On one occasion, it was interesting to hear the following dialogue of two students:

NANCY: I understand the rule so I don't see a reason why it shouldn't be correct.
JANET: Yeah. I understand yours.
RESEARCHER: So you don't see any reason why it's not correct either?
JANET: No.

Mainly Nancy, but also Janet influenced by her partner, failed to see the need to justify their rules. Their correctness was so obvious in their minds and so "understandable" that justification seemed unnecessary. In these cases, the relationship between the generic and general seems to be self-evident, thanks to the relationship between construction and expression: the situated abstraction goes something like this – "if you build it like that, and you say what you've built, it must be right". Such an approach was characterised as *authoritarian justification* (number of responses: 3). It is worth mentioning though that after their initial reaction to use the authoritarian strategy, as their discussions continued, they used other strategies too. The same outcome holds for Emily and Maria, as will be presented below in the summative results in Table 6.2.

Summative Results

In Table 6.2, we summarise the results by presenting the different strategies students used in their pairs. Each column that corresponds to a strategy is split into two cells to distinguish through the use of a tick whether the student named first or the one named second in the pair or both students used that strategy. As mentioned earlier, some students used more than one strategy. For example, Emily and Patricia were two

Table 6.2 Justification Strategies for Rules Correctness Used by Each Pair of Students from Each School. A Tick Means That the Correspondent Student Has Used That Strategy. For Example, Emily Used Both Constructive and Empirical, Whereas, Patricia, Only Structural Justification

	Students	Constructive		Structural		Empirical		Authoritarian	
School A (Yr 7)	Emily + Patricia	3			3	3			
	Alex + Anne	3	3						
	Janet + Nancy	3			3			3	3
School B (Yr 7)	Alan + Simon			3	3				
	Fiona + Jackie	3	3	3	3				
	Susan + Dorothy	3	3	3	3				
School B (Yr 8)	Neil + Rex	3	3			3			
	Randy + Susan	3					3		
	Tod + Ally			3	3				
School C (Yr 8)	Colin + Lara			3	3				
	Maria + Mike	3	3	3	3				
	Penny + Leo			3	3				
	Alicia + Greg			3	3				
	Abigail + Mark	3	3						
	Scot + Louise	3	3	3	3				
	Amy + Nick			3	3				
	Eleanor + Trevor			3	3				
School C (Yr 9)	Andy + Penny	3	3	3					
	Bill + Dave	3			3	3			
	Carey + Teddy	3				3	3		
	Nick + Scot			3	3	3			
	Colin + Maria	3			3				3
	Eleanor + Mark	3	3			3			
	Greg + Leo	3	3			3	3		
TOTAL		16	10	13	16	7	3	1	2
		26		29		10		3	

Year 7 students from school A. Emily used the constructive and the empirical strategies, whereas Patricia used the structural one only.

The last row in Table 6.2 reveals the total number of students who used each strategy. Out of the 68 times all strategies were used, the constructive strategy was used 26 times and the structural 29. Thus a justification strategy relevant to the structure of the constructed model was used to support the rules' correctness 55 times. In total, about four-fifths of students' strategies relied on the structure of the model.

Given the size of the sample, we do not of course claim causality. However, we do conjecture that there could be some link to the main Train-Track activity students had worked on prior to their collaborative activity. It seems that the system's design to encourage students to focus on the structure prevailed against their tendency to focus on recursive rather than functional relationships, a strongly prevalent tendency revealed in other studies (e.g., Lee, 1996; Stacey, 1989). Even though we noticed that a few students (2 Year 7 and 1 Year 8) followed the empirical strategy initially, we can see how their further interaction with eXpresser and their constructed models during their collaboration shaped their thinking in a direction that takes into account the structure of the pattern.

Justification for Equivalence

During the second part of the collaborative activity, students in their same pairs compared each other's rules and discussed their equivalence. They were asked: *Can you explain together why your rules look different but are equivalent? Discuss and write your explanations.*

After extensive analysis, coding and recoding, the data ended up grouped into three main categories: *structural, symbolic,* and *empirical,* described below along with their subcategories. Note that 22 out of the 48 students used more than one strategy to justify the rules' equivalence.

Structural Justification for Equivalence

Justifications in this category all focused on the structural aspect of the pattern by, for example, comparing the building blocks used in the different patterns and making arguments as to their equivalence with little if any reference to the symbolic rule. We distinguished three subcategories illustrated below with data from the study.

1. Reconstructive Justification (number of responses: *30*). In this subcategory, different building blocks are compared or reconfigured as illustrated by the case of Janet and Nancy (see Figure 6.4).

Models	Janet	Nancy
Building Blocks	Red $=$ $\boxed{4} \times$ Blue $=$ $\boxed{3} \times$ Green $=$ $\boxed{2} \times$	Green $=$ $\boxed{7} \times$ Blue $=$ $\boxed{5} \times$
eXpresser Rule	$\boxed{4} \cdot \frac{\text{number of red fills}}{5} + \boxed{3} \cdot \frac{\text{number of red fills}}{5} + \boxed{1} + \boxed{2} \cdot \boxed{5}$	$\frac{\text{NumOf}}{5} \times \boxed{7} + \boxed{5} \times \boxed{1}$
Algebraic Rule	$4n + 3(n{+}1) + 2{\times}1$	$7n + 5{\times}1$

Figure 6.4 Janet and Nancy's model, building blocks and general rules.

Nancy compared her building block with that of Janet's:

NANCY: Yeah it's one red building block plus one blue building block so that would actually kind of make the...

JANET: Yeah, it would make the same shape...

NANCY: Because one red building block added to one blue building block...

JANET: And that's the same as one of my green building blocks.

Students complemented each other's arguments and concluded that their building blocks were the same. They either explicitly related the models to their rules or linked the number of tiles in each block to the coefficients in the algebraic expressions. Rather, they simply compared the building blocks underlying the patterns used, basing their verbal interactions on their shared experience of construction and (algebraic) expression.

2. Experimental Justification (number of responses: 8). In this sub-category, students chose a specific case and compared their two models and rules for this case, as illustrated by Alex and Anne (see Figure 6.5).

ALEX: I kind of got a C, but coloured them in different ways so I mean the 5 is only added at the end...

ANNE: Then there are just 7 tiles in one model.

ALEX: Yes, but your first model has 12 tiles and your second model has 7 tiles. For 5 red blocks I have 5 blue extra tiles, but you have 12 blue extra tiles.

Anne was able to read Alex's rule and recognised the configuration of tiles that formed a similar building block to hers. Yet it was

Models	Alex		Anne	
Building Blocks	Green $= \boxed{2} \times$ ▨		Blue ▤ $= \boxed{12} \times$ ▢	
	Red ▤ $= \boxed{3} \times$ ▢			
			Green ⌉ $= \boxed{7} \times$ ▨	
	Blue ▤ $= \boxed{5} \times$ ▢			
eXpresser Rule	[number of red tiles / 5] $\times \boxed{2} + \boxed{2} + \boxed{5}$ ... [number of red tiles] $\times \boxed{3} + \boxed{5}$		$\boxed{7} \times$ [As above number / 5] $+ \boxed{12}$	
Algebraic Rule	$(2n \times 2) \times 2 - 3n - 5$		$7n + 12$	

Figure 6.5 Alex's and Anne's model, building blocks, and general rules.

evident that both students considered each building block as a separate model. At first, Alex chose to change the number of red blocks in her model to 5 to match Anne's model, but then realised that it was just not possible to match: the two models, in fact, had different constant terms. Alex decided to compare the two models for the *same* model number and then justified the non-equivalence of the two rules.

This strategy was used by six other students, who all chose to select a value for their model number and use it in both rules (their own and their pair's one), but then compared their models structurally focusing mainly on their length, but also on their building blocks. Such a strategy is mathematically valid for justifying non-equivalence (as demonstrated by Alex and Anne's example above), since finding one case for which two rules are not equivalent is enough to disprove equivalence. However, experimenting with one case is not enough to generalise the equivalence for any case.

3. Justification by Contradiction (number of responses: 7). Students used the same model number and calculated the number of tiles used, and noticed that they obtain different answers, as illustrated by Amy and Nick (Figure 6.6). They had to go back to comparing their models structurally.

Nick noticed that for the same value of the independent variable, their models could *never* be the same. His justification was based on a contradiction expressed both numerically and structurally.

There were five more students that used justification by contradiction and they all resorted to compare the structure of their models.

Models	Amy		Nick	
Building Blocks	Red $\quad$ = $\boxed{5}$ × $\square$		Green $\quad$ = $\boxed{7}$ ×	
	Yellow $\quad$ = $\boxed{9}$ ×		Yellow $\quad$ = $\boxed{5}$ ×	
eXpresser Rule	$\boxed{5}$ × $\boxed{5}$ + $\boxed{2}$ × $\boxed{9}$		$\boxed{5}$ × $\boxed{7}$ + $\boxed{5}$	
Algebraic Rule	$5n + 2 \times 9$		$7n + 5$	

Figure 6.6 Amy's and Nick's model, building blocks and general rules.

Symbolic Justification (Number of Responses: 21)

This category comprised student justifications focused on their eXresser rules and justified their equivalence by adding the constants and variables in each rule and comparing them as illustrated by Leo and Penny's case (Figure 6.7).

When paired, Leo realised that his rule was incorrect, but was able to derive a correct general rule that he wrote on paper as $[5] \times 9 - [5] \times 2 + 5$. This is what they both compared with Penny's rule.

LEO: I had 5 times 9 because I had 9 things but I have to take away 2 of my red building block, so I have to take away 10 tiles because I need to have 5 sevens. I had that many on the end of each one [pointing at his model]. That is why I have to take away 2 and then plus 5 because I need an extra line at the end. The 9 minus 2 is equal to plus 7 and the 5 is the same and then the 5 is the same so they're the same rule but written differently.

PENNY: Mine is 5 times 1 plus 8 times 7. These 8 times 7 because we've got 8 of the 7 blocks and so 8 times 9 minus times 2 is 8 times 7.

They concluded that Leo's second rule on paper was equivalent to Penny's rule. In this example, the value of reflecting on one's own rule, triggered by collaboration is revealed.

There are 19 more students that followed this strategy and as shown in Table 6.3, the majority were Year 9 students. This might be due to their greater experience compared to Year 7 and 8 students with algebraic language. Two Year 9 students, Eleanor and Carey, resorted initially to the symbolic justification (see Table 6.3) and then used some type of a structural justification to support further their arguments and visually justify their rules' equivalence or non-equivalence. The

Models	Penny		Leo	
Building Blocks	Green	$= \boxed{5} \times \blacksquare$	Red	$= \boxed{9} \times \square$
	Yellow	$= \boxed{7} \times \blacksquare$		
eXpresser Rule	$\boxed{5} \times \boxed{1} + \boxed{8} \times \boxed{7}$		$\boxed{5} \times \boxed{9}$	
Algebraic Rule	$5 \times 1 + 7n$		$n + 9$	

Figure 6.7 Penny's and Leo's model, building blocks, and general rules.

Table 6.3 Justification Strategies for Rules Equivalence Used by Each Pair of Students from Each School. Similarly to Table 6.2, a Tick Means that the Correspondent Student Has Used That Strategy. For Example, Emily Used Only the Symbolic Justification Strategies, Whereas Her Partner, Patricia, Used Only the Reconstructive Justification Strategy

	Students	Rec.		Exper.		Contr.		Sym.		Match		Eval.	
School A (Yr 7)	Emily + Patricia		3					3					
	Alex + Anne	3	3	3	3	3	3			3		3	
	Janet + Nancy	3	3					3					
School B (Yr 7)	Alan + Simon			3	3					3	3		
	Fiona + Jackie	3	3					3	3				
	Susan + Dorothy	3	3					3	3				
School B (Yr 8)	Neil + Rex			3	3								
	Randy + Susan	3	3										
	Tod + Ally	3	3										
School C (Yr 8)	Colin + Lara	3	3										
	Maria + Mike							3	3				
	Penny + Leo			3				3	3				
	Alicia + Greg	3	3										
	Abigail + Mark	3	3			3	3						
	Scot + Louise	3	3			3							
	Amy + Nick					3	3	3			3	3	
	Eleanor + Trevor	3	3					3					
School C (Yr 9)	Andy + Penny							3	3	3			
	Bill + Dave	3						3					
	Carey + Teddy	3						3				3	3
	Nick + Scot	3	3								3		
	Colin + Maria							3	3				
	Eleanor + Mark			3		3	3	3					
	Greg + Leo							3		3			3
		13	17	4	4	4	3	13	8	3	2	3	2
TOTAL		30		8		7		21		5		5	

rest of the students started off structurally and then moved on to symbolic justification.

Empirical Justification for Equivalence

Some students focused solely on the numerical aspect of the rules, avoiding any reference to the structure of their model constructions. Two subcategories were distinguished:

1. Matching-Terms Justification (number of responses: 5). In this category students pick a constant or a variable and compare with the

equivalent term in the other students' rules. Here is Alex at an early stage of her collaboration with Anne:

ALEX: They both have 7 in them plus something to make the end of the pattern.

She picked a constant in her rule and identified it in Anne's rule too (see Figure 6.10). She noticed the similarities in the algebraic expressions, but also the difference in the added constant term (5 in Alex's rule, but 12 in Anne's rule). This initial reaction to the collaborative activity reveals their tendency for an exploration of the rules, a rather important problem-solving skill, and does allow them to "read" both their rules and identify commonalities and differences.

Only two Year 9 students, Andy and Scot, as presented in Table 6.3, used this strategy after they used a symbolic and a structural justification strategy respectively. It seems that this helped them to simplify further the rules and prove that their rules were equivalent and identified their simplest form as 7n+5. Similarly, two Year 7 students, Alan and Simon, as also presented in Table 6.3, used the matching-terms strategy to support further their arguments after using the experimental justification strategy. Both these strategies revealed those students' tendency to compare models first and then rules for specific cases preventing them from focusing on the general case and their rules' equivalence for any value for the model number. Alex, who was the other Year 7 student, used this strategy too. She seemed eager to use many different arguments to support her view on her pair's rule not being equivalent to hers and she was the only student to use five different strategies.

2. Evaluating-Terms Justification (number of responses: 5). In this category, students compared the number of tiles for different model numbers. Later in their discussion, Alex chose a value for the independent variable and compared the answers for the two rules:

ALEX: Model number 1 is blue blocks and it's got 12 tiles in total. The backwards C is model number 2. So, we have 12 plus 7...19 tiles.
ANNE: No, model number 2 is 2 backward Cs plus the blue block. So, 2 times 7 plus 12...26 tiles.

Anne's answer included seven more tiles because of the blue block she had added to her model. The students were confused at this point as to what the model was and what the model number was. This was the last strategy Alex used to support her argument on non-equivalence of their rules.

All three Year 9 students who used this strategy, except for Teddy, which was his only strategy, used it as an additional justification strategy, but focused mostly on their original one, which was the symbolic strategy. A similar story holds for the only Year 8 student, Amy, who used the evaluating-terms justification strategy.

Summative Results

In Table 6.3, we summarise the results by presenting the different strategies students used in their pairs to justify their rules' equivalence or non-equivalence. The same structure as for Table 6.2 is used. This means that each column that corresponds to a strategy is split into two cells to distinguish through the use of a "tick" whether the student named first or the one named second in the pair or both students used that strategy.

The last row in Table 6.3 reveals the total number of students who used each strategy. Out of the 76 times all strategies were used, a type of the structural justification strategy was used 45 times with the reconstructive strategy dominating (30 times) and the symbolic one 21 times: the empirical strategy was the least used (10 times). About three-fifths of the time, students' strategies relied on the structure of the model and two-fifths on the rules alone through manipulation of the terms. Compared to the first part of the collaborative activity on discussing correctness, students seem to rely mostly on the way they constructed their models. Some of the students, the majority of whom were in Year 9, focused on manipulating their rules algebraically to transform them into their simplest form. These students revealed their confidence in using the eXpresser language but crucially related it to formal algebraic language identifying correctly and manipulating successfully constants and variables. Despite the obvious limitations of this quantitative overview, it suggests that a focus on structure remained the dominant choice for students. This, in combination with the general tendency in the literature and our previous anecdotal observations and evidence that seemed to suggest a prevalence of empirical justifications, supports the design decisions of the eXpresser and how it can act as a context for the collaboration activities.

Discussion

The data in the previous section support the value of reflective and collaborative activities. Similarly to many researchers we mentioned earlier, such as Ellis (2011), Wood (1988) and Lou et al. (1996), students engaged in acts of argumentation to support their own solution strategies, but also recognise and understand those of their peers. Such actions enhanced their justification skills and their algebraic thinking. For example, there were many cases, both during individual and initial collaborative work, that manifested students' particular misconceptions such as their tendency to focus on the additive principle as has been reported before (e.g., Hart, 1981; Lee, 1996; Stacey, 1989) or what the general rule is and what "n" represents. In our case, the crucial point is that the collaborative activity helped students to share their misunderstandings and support each other to overcome them *because they had an object to share*.

Besides challenging misconceptions about generality, putting students in a collaborative setup to discuss their modelling approaches helped them reflect on the components of their models in relation to the corresponding rules. From judging the justification strategies students used, it seemed that most students made sense of the "unknown" as they were given a rationale for its use (c.f. Filloy et al., 2008). Also, students were able to manipulate the rules and avoid mistakes, such as $Ax + B = (A + B)x$, as they could easily identify the components in their models' rules.

In terms of forming heterogeneous groups, as suggested by Leonard (2001), our first criterion was for students to have derived *different* rules for the argumentation activity to be meaningful. The second criterion, the students' characters and which pairs would collaborate sensibly and constructively, relied a lot upon the teacher who advised us based on the suggested pairs by the Grouping Tool. Students were prompted by the activity questions and also the researchers and the teacher reminded them on a regular basis to share arguments and consider all their different approaches. This established a collaborative learning approach as emphasized by Lou et al. (1996) and promoted a collaborative "knowledge building" culture as described by Scardamalia and Bereiter (2006), which led students to use more than one justification strategy, resolve contradictions, and therefore challenged and broadened their algebraic ways of thinking.

Revisiting Dretske's (1990) distinction between sensory and cognitive mode as two types for the act of coming to visually perceive a pattern, it can be argued that the distinction between the two relies crucially on the tools for expression the student is given as well as the context in which they are used. In the case of MiGen, it can be argued that when students interacted individually with the eXpresser, they could have seen patterns and their components as mere objects (sensory) and they could have derived expressions to "link" them procedurally. However, when they became involved in the collaborative activity, they were encouraged to reflect upon these objects, recognise their properties, and argue about their expressions-rules (cognitive). Even though eXpresser is designed to support students' expressions of generalisations, we argue that it is students' engagement in acts of justification through the collaborative activity that supported their understanding and "forced" them to reflect upon their visual perception of the figural patterns.

Since students are usually presented with the method of generalising based on noticing a commonality among the terms in a pattern sequence, one outstanding challenge posed by some researchers is to understand how students come to generalise what they notice to the *whole pattern* (e.g., Radford, 2010). In eXpresser, this turns out to be relatively straightforward, as with eXpresser's functionality of building a pattern using a building block, the arbitrary repetition of this building block when the model is, animated, shows how the model is "generalised". We found that structural and reconstructive justifications were the main strategies used, a result that seems to align with students' intuitive explanation of their construction method. As also claimed by Ellis (2007), Küchemann (2010), and Arcavi et al. (2017), attending to the structure of a pattern has a better chance of justifying students' expressions generally and the findings from our studies seem to support this claim as the students successfully reached a good understanding of generality as demonstrated by their justification strategies.

The choice of a "unit of repetition" or (in the eXpresser's language) a "building block" proved to be a crucial step towards a correct generalisation, as suggested by Rivera (2010). Supported by the intelligent support of the MiGen system, students could find the minimum number of tiles that could be grouped into a building block and then repeated to form a pattern. Those who didn't were challenged by their pair during collaboration (see, for example, the case of Anne, who chose 12 tiles for her blue building block instead of five, and was challenged by Alex), an outcome that emphasises the value of collaboration towards forming correct generalisations.

Students' investment in building their own models supported them in deriving generalisations by directing their focus towards relationships between quantities. The algebraic discourse of the eXpresser – the grammar of objects and relationships between them – gave students a means to express generalisation without the formal machinery of algebra. We argue that students were supported in expressing a general rule by the eXpresser's language that gave meaning to the 'unknown', allowing them to name it as it made sense to them and use this as an intermediate step to formal algebraic language. Therefore, we supported the transition from natural language, as "always" and "every", which is more intuitive for students (e.g., Warren & Cooper, 2008), to formal algebraic language. This action, which we refer to as "expressive action" in Figure 6.1 and which could be linked to the "symbolic action" described by Rivera (2010), was supported further by the collaborative activity when students were encouraged to reflect upon their expressions and endorse them.

In more detail, during the collaboration phase, students were encouraged to revise their rules and identify the constants and the variables. The metaphor of the unlocked number was particularly salient here. Perhaps surprisingly all the students realised that in this case this was also the model number, "the number that can change" or in other words the *variable*.

An interesting observation is that some students who used algebraic symbolic justification responded in ways that could allow us to assume that the transition to the use of letters to represent the unknown seemed easier for them. For example, Penny, during collaboration:

RESEARCHER: How many tiles in model number 6?
PENNY: So it will be 5 times 1 plus x times 7.
RESEARCHER: And what's x?
PENNY: x is 6.

Although it is too early to claim that this is the case and despite lack of specific data on this question directly, we can surmise based on the above and similar interchanges that students mostly seemed to understand what the "n" stands for and equally important, convinced of a rationale for having a general rule. Out of all year groups, most students (nine in total) who used a symbolic justification strategy were Year 9 students (six were Year 7 and six were Year 8), which could mean that they are more experienced with algebraic notation or they have gained more expertise in the use of eXpresser.

The difference between algebraic thinking and algebraic symbolism is evident in these data. Students were able to express generality verbally and in written form. If we consider the trajectory, there is a change from the first time they interacted with the eXpresser and their latest interactions in terms of their expressions of generality. By the end, most students were able to write down the rules using the eXpresser language (not the boxes, but e.g., $5 \times$ Model Number $+ 3$, or $5 \times$ Unlocked Number $+ 3$).

What is encouraging is that most students' activities after the engagement described here tended to continue to focus on the structure of the patterns in order to articulate the general rule. In their efforts to justify their general rules, students revisited their generalising actions, built on them, and constructed ones that were more powerful and meaningful. They succeeded in reaching rich justifications for the correctness and equivalence of their derived algebraic expressions for the linear Train-Track pattern.

The findings point to the students' preference for referring to the structure of their models to justify correctness of their rules, since most students (55 out of 68 times all strategies were used or 81%) used either the constructive or the structural justification strategies. There was a similar preference when students justified the equivalence of their rules, since most students (45 out of 76 times all strategies were used or 59%) used a type of the structural justification strategy. The second most common strategy (21 out of 76 times all strategies were used or 39%) in terms of equivalence was symbolic justification. This result supports the usefulness of the eXpresser for students' introduction to algebra and the collaborative activity for a possible introduction to proof (as a next step from justification).

In summary, we can claim that students' engagement in acts of justifying through collaboration seemed to support their generalisation skills in a number of ways that is: recognise the importance of seeing the structure, find the constants and the variables in their model and rule, express relationships using an independent variable to link patterns within their models and see the rationale and recognise the power of structural sense and algebraic generalisation. But to achieve this they had engaged in a carefully designed sequence of activities in the context of the eXpresser, which we suggest played a key role in their learning outcomes and assisted the integration of transition from arithmetic to algebra.

Acknowledgements

Early stages of the analysis of the data discussed in this chapter were presented at the 35th Conference of the International Group for the Psychology of Mathematics Education, 10–15 July 2011, in Ankara, Turkey. The reader is referred to Geraniou, E., Mavrikis, M., Hoyles, C. and Noss, R. (2011). "Students" justification strategies on equivalence of quasi-algebraic expressions', *PME35: International Group for the Psychology of Mathematics Education,* Vol. 2, pp.393–400. 10th–15th of July 2011, Ankara, Turkey. We would like also to thank Professors Celia Hoyles and Richard Noss for their leadership in the MiGen project as well as the rest of the MiGen team for their support and ideas in the work presented here.

Notes

1 For this and other design priorities and rationales, the reader is referred to Noss et al. (2012). Mavrikis et al. (2013b) provides a more detailed discussion on how the design of eXpresser supports the development of algebraic ways of thinking.

2 To help the reader, the number of white gaps inside the Train-Track model could be mapped to the value of the Model Number.

3 Apart from immediate feedback from the visualisations and representations of the microworld (e.g., lack of colouring of a pattern) the MiGen system incorporates intelligent components that analyse students' activities and provide explicit feedback on their actions. This involves nudges to draw students' attention to inconsistencies in their model compared to the activity model, for example, the lack of colouring or structural generality of the pattern and other prompts to help them reflect explicitly on their actions, especially when they request additional help. For more details and examples, see Noss et al. (2012) and Mavrikis et al. (2013a).

4 All numbers in eXpresser are *constants* by default, referred to as "locked" numbers. When the user "unlocks a number", it is possible to change its value; it becomes a *variable*.

5 For more information on the Grouping Tool, see Noss et al. (2012) and Gutierrez-Santos et al. (2017).

6 The collaborative activity, though, is designed to be carried out in the classroom where the teacher is expected to run a classroom discussion at the end of the lesson.

7 http://www.transana.org/

8 When interacting with students, a pedagogical strategy, referred to as "messing-up" (Healy et al., 1994) was used. This strategy challenges students to construct models that are impervious to changing values of the various properties of their construction.

References

Abdu, R., Olsher, S., & Yerushalmy, M. Towards automated grouping: Unraveling mathematics teachers' considerations. In Barzel, B., Berbinik, R., Göbel, L., Pohl, M., Ruchniewicz, H., Schacht, F., & Thurm, D. (Eds.) *Proceedings of the 14th International Conference on Technology in Mathematics Teaching-ICTMT 14* (Essen, Germany, July 2019). [S.147] https://duepublico2.uni-due.de/receive/duepublico_mods_00048820

Arcavi, A., Drijvers, P., & Stacey, K. (2017). *The learning and teaching of algebra: Ideas, Insights and Activities.* New York: Routledge.

Bereiter, C., & Scardamalia, M. (2003). Learning to work creatively with knowledge. In E. D. Corte, L. Verschaffel, N. Entwistle, & J. V. Merriënboer (Eds.), *Powerful learning environments: Unravelling basic components and dimensions* (pp. 73–78). Oxford: Elsevier Science.

Bills, L., & Rowland, T. (1999). Examples, generalisation and proof. In L. Brown (ed.), *Making meanings in mathematics* (pp. 103–11). York: QED.

Carter, G., & Jones, G. M. (1993). *The relationship between ability-paired interactions and the development of fifth graders' concepts of balance.* Paper presented at the Conference of the National Association of Research in Science Teaching: Atlanta. (ERIC Document Reproduction Service No. ED 361 175).

Chua, B. L., & Hoyles, C. (2011). Secondary school students' perception of best help generalising strategies. *Paper presented at the Seventh Congress of the European Society for Research in Mathematics Education.* Rzeszów, Poland.

Coe, R., & Ruthven, K. (1994). Proof practices and constructs of advanced mathematics students. *British Educational Research Journal, 20*(1), 41–53.

Cohen, E. G., & Lotan, R. A. (1995). Producing equal-status interaction in the heterogeneous classroom. *American Educational Research Journal, 32*(1) resulta), 99–120.

Dörfler, W. (2008). En route from patterns to algebra: Comments and reflections. *ZDM – The International Journal on Mathematics Education, 40*(1), 143–160.

Dretske, F. (1990). Seeing, believing, and knowing. In D. Osherson, S. M. Kosslyn, & J. Hollerback (Eds.), *Visual cognition and action: An invitation to cognitive science* (pp. 129–148). Cambridge, Massachusetts: MIT Press.

Ellis, A. B. (2007). Connections between generalising and justifying: students' reasoning with linear relationships. *Journal for Research in Mathematics Education, 38*(3), 194–229.

Ellis, A. B. (2011). Generalising-promoting actions: How classroom collaborations can support students' mathematical generalizations. *Journal for Research in Mathematics Education, 42*(4), 308–341.

Filloy, E., & Rojano, T. (1989). Solving equations: The transition from arithmetic to algebra. *For the Learning of Mathematics*, 9(2), 12–25.

Filloy, E., Rojano, T., & Puig, L. (2008). *Educational algebra; a theoretical and empirical approach*. Springer: New York.

Filloy, E., Rojano, T., & Solares, A. (2010). Problems dealing with unknown quantities and two different levels of representing unknowns. *Journal for Research in Mathematics Education*, 41(1), 52–80.

Fuchs, L., Fuchs, D., Karns, K., Hamlett, C., Katzaroff, C., & Dutka, S. (1998). Comparisons among individual and cooperative performance assessments and other measures of mathematics competence. *The Elementary School Journal*, 99(1), 23–52.

Geraniou, E., Mavrikis, M., Kahn, K., Hoyles, C., & Noss, R. (2009). *Developing a Microworld to Support Mathematical Generalisation. Proceedings of the 33rd Conference of the International Group for the Psychology of Mathematics Education* (Vol. 3, 49–56). Thessaloniki, Greece.

Gutierrez-Santos, S., Mavrikis, M., Geraniou, E., & Poulovassilis, A. (2017). Similarity-based grouping to support teachers on collaborative activities in an exploratory mathematical microworld. *IEEE Transactions on Emerging Topics in Computing*, 5(1), 56–68.

Harel, I., & Papert, S. (1991). *Constructionism*. Norwood, N.J.: Ablex Publishing Corporation.

Harel, G., & Tall, D. (1991). The general, the abstract and the generic in advanced mathematics. *For the Learning of Mathematics*, 11(1), 38–42.

Hart, K. (1981) Ratio and proportion. In *Children's understanding of mathematics 11–16*, edited by Kathleen Hart, 88–101. London: John Murray.

Healy, L., & Hoyles, C. (2000). A study of proof conceptions in algebra. *Journal for Research in Mathematics Education*, 31(4), 396–428.

Healy, L., Pozzi, S., & Hoyles, C. (1995). Making sense of groups, computers and mathematics. *Cognition and Instruction*, 13(4), 505–523.

Healy, L., Hoelzl, R., Hoyles, C., & Noss, R. (1994). Messing up. *Micromath*, 10(1), 14–17.

Kaput, J. (1999). Teaching and learning a new algebra. In T. Romberg & E. Fennema (Eds.), *Mathematics classrooms that promote understanding* (pp. 133–155). Hillsdale, NJ: Lawrence Erlbaum.

Kieran, C. (1989). The early learning of algebra: A structural perspective. In S. Wagner, & C. Kieran (Eds.), *Research Issues in the learning and teaching of algebra*. (pp.33–56). VA: LEA.

Kieran, C. (2001). The mathematical discourse of 13-year-old partnered problem solving and its relation to the mathematics that emerges. *Educational Studies in Mathematics*, 46(13), 187–228.

Küchemann, D., & Hoyles, C. (2009). From empirical to structural reasoning in mathematics: Tracking changes over time. In D. Stylianou, M. Blanton & E. Knuth (Eds.), *Teaching and learning proof across the grades K-16 perspective* (pp. 171–191). London: LEA.

Küchemann, D. (2010). Using patterns generically to see structure. *Pedagogies: An International Journal*, 5(3), 233–250.

Laborde, C. (1994). Working in small groups: A learning situation? In R. Beiler (Ed.), *Didactics of mathematics as a scientific discipline* (pp. 147–157). Dordrecht: Kluwer.

Lavy, I., & Leron, U. (2004). The emergence of mathematical collaboration in an interactive computer environment. *International Journal of Computers for Mathematical Learning*, 9(1), 1–23.

Lee, L. (1996). An initiation to algebraic culture through generalization activities. In N. Bednarz, C. Kieran & L. Lee (Eds.), *Approaches to algebra. Perspectives for research and teaching* (pp. 87–106). Dordrecht, The Netherlands: Kluwer Academic.

Leonard, J. (2001). How group composition influenced the achievement of sixth-grade mathematics students. *Mathematical Thinking and Learning*, 3(2&3), 175–200.

Linchevski, L., & Kutscher, B. (1998). Tell me with whom you're learning, and I'll tell you how much you've learned: Mixed-ability versus same-ability grouping in mathematics. *Journal for Research in Mathematics Education*, 29(5), 533–554.

Lou, Y., Abrami, P., Spence, J., Poulsen, C., Chambers, B., & D'Apollonia, S. (1996). Within-class grouping: A meta-analysis. *Review of Educational Research*, 66(4), 423–458.

Mason, J. (1996). Expressing generality and roots of algebra. In N. Bednarz, C. Kieran, & L. Lee (Eds.), *Approaches to algebra – perspectives for research and teaching* (pp. 65–86). The Netherlands: Kluwer Academic Publishers.

Mavrikis, M., Gutierrez-Santos, S., Geraniou, E., & Noss, R. (2013). Design requirements, student perception indicators and validation metrics for intelligent exploratory learning environments. *Personal and Ubiquitous Computing*, 17(8), 1605–1620. doi: http://dx.doi.org/10.1007/s00779-012-0524-3.

Mavrikis, M., Noss, R., Geraniou, E., & Hoyles, C. (2013b). Sowing the seeds of algebraic generalisation: Designing epistemic affordances for an intelligent microworld. In R. Noss & A. DiSessa (Eds.), *Special issue on knowledge transformation, design and technology, journal of computer assisted learning*, 29(1), 68–85. doi: 10.1111/j.1365-2729.2011.00469.x

Moss, J., & Beatty, R. (2006). Knowledge building in mathematics: Supporting collaborative learning in pattern problems. *Computer-Supported Collaborative Learning*, 1(4), 441–465.

Mercer, N. (1995). *The guided construction of knowledge*. Clevendon: Multilingual Matters.

Mercer, N. (2000). *Words and Minds: how we use language to think together*. London: Routledge.

Mor, Y., Tholander, J., & Holmberg, J. (2005). Designing for cross-cultural web-based knowledge building. In Timothy Koschmann, Daniel D. Suthers, & Tak-Wai Chan (eds.), *"The 10th Computer Supported Collaborative Learning (CSCL) Conference (2005)"* (pp. 450–459). Lawrence Erlbaum Associates, Taipei, Taiwan.

Mor, Y., Noss, R., Hoyles, C., Kahn, K., & Simpson, G. (2006). Designing to see and share structure in number sequences. *International Journal for Technology in Mathematics Education*, 13(2), 65–78.

Noss, R., Healy, L., & Hoyles, C. (1997). The construction of mathematical meanings: Connecting the visual with the symbolic. *Educational Studies in Mathematics*, 33(2), 203–233.

Noss, R., & Hoyles, C. (1996). *Windows on mathematical meanings: Learning cultures and computers*. Dordrecht: Kluwer Academic Publishers.

Noss, R., Hoyles, C., Mavrikis, M., Geraniou, E., Gutierrez-Santos, S., & Pearce, D. (2009). Broadening the sense of 'dynamic': A microworld to support students' mathematical generalisation. *ZDM, 41*(4), 493–503.

Noss, R., Poulovassilis, A., Geraniou, E., Gutierrez-Santos, S., Hoyles, C., Kahn, K., Magoulas, G., & Mavrikis, M. (2012). The design of a system to support exploratory learning of algebraic generalisation. *Computers and Education, 59*(1), 63–81.

Radford, L. (2009). Why do gestures matter? Sensuous cognition and the palpability of mathematical meanings. *Educational Studies in Mathematics, 70*(2), 111–126.

Radford, L. (2010). Layers of generality and types of generalization in pattern activities. *PNA-Pensamiento Númerico Avanzado, 4*(2), 37–62.

Radford, L. (2011). Grade 2 students' non-symbolic algebraic thinking. In J. Cai & E. Knuth (Eds.), *Early algebraization* (pp. 303–322). Berlin: Springer-Verlag.

Radford, L., Bardini, C., & Sabena, C. (2007). Perceiving the general. The multi-semiotic dimension of students' algebraic activity. *Journal for Research in Mathematics Education, 28*(5), 507–530.

Rivera, F., & Becker, J. (2008). Middle school children's cognitive perceptions of constructive and deconstructive generalizations involving linear and figural patterns. *ZDM, 40*(1), 65–82.

Rivera, F. (2010). Visual templates in pattern generalization activity. *Educational Studies in Mathematics, 73*(3), 297–328.

Scardamalia, M., & Bereiter, C. (2006). Knowledge building: Theory, pedagogy, and technology. In K. Sawyer (Ed.), *Cambridge handbook of the learning sciences* (pp. 97–118). New York: Cambridge University Press.

Stacey, K. (1989). Finding and using patterns in linear generalizing problems. *Educational Studies in Mathematics, 20*(2), 147–164.

Strauss, A., & Corbin, J. (1998). *Basics of qualitative research: techniques and procedures for developing grounded theory.* London: Sage.

Swan, M. (2006). *Collaborative learning in mathematics: A challenge to our beliefs and practices.* London: National Institute for Advanced and Continuing Education (NIACE); National Research and Development Centre for Adult Literacy and Numeracy (NRDC).

Teasley, S. D., & Roschelle, J. (1993). Constructing a joint problem space: The computer as a tool for sharing knowledge. In S. P. Lajoie & S. J. Derry (Eds.), *Computers as cognitive tools* (pp. 229–258). Hillsdale, NJ: Lawrence Erlbaum Associates, Inc.

Vahey, P., Enyedy, N., & Gifford, B. (2000). Learning probability through the use of a collaborative, inquiry-based simulation environment. *Journal of Interactive Learning Research, 11*(1), 51–84. Charlottesville, VA: AACE.

Vahey, P., Tatar, D., & Roschelle, J. (2007). Using handheld technology to move between private and public interactions in the classroom. In M. van 't Hooft & K. Swan (Eds.), *Ubiquitous computing in education: Invisible technology, visible impact* (pp. 187–210). Mahway, NJ: Lawrence Erlbaum Associates.

Warren, E., & Cooper, T. (2008). Generalising the pattern rule for visual growth patterns: Actions that support 8 year olds' thinking. *Educational Studies in Mathematics, 67*(2), 171–185.

Wood, D. (1988). *How children think and learn.* Oxford and Cambridge, MA: Blackwell.

7 The Importance of Algebra Structure Sense for the Teaching and Learning of Mathematics

Teresa Rojano and Santiago Palmas

Mathematical capabilities from different educational levels that are directly related to development of algebra structure sense are made explicit and arguments are presented that support the idea that this development contributes to the mathematical maturity of students. Emphasis is placed on the importance of teachers and designers of curriculum and teaching materials (digital or otherwise) incorporating into their practice elements that foster development of a sense of the structure in students.

Introduction

Kieran (2004) classifies algebraic activity as generative (related to the algebraic interpretation and representation of situations, properties, patterns, and relationships); transformative (relating to all types of symbolic manipulation); and global/meta-level (involving more general mathematical processes related to contexts than motivate the use of algebra). Here we will use the term transformational algebra, coined by Kieran, in its broadest sense, that is, that it encompasses not only symbolic skills but also includes conceptual elements (Kieran, 2013). A section below sets out the author's arguments, contending that manipulative algebra has this dual nature. This chapter makes explicit aspects of transformational algebra that are related to the development of algebra structure sense (ASS), and advances arguments that support the idea that the latter contributes to students' mathematical maturity, either in its role as a prerequisite for the study of post-secondary school topics (algebraic or others) or as prior capability to access topics of advanced mathematics.

In the next section, transformative algebra topics in the mathematics curriculum are found and those that include ASS aspects are identified. To that end, reference is made to curricular documents from four countries: the USA, UK, South Korea, and Mexico. Moreover, we present examples that illustrate the role of ASS in teaching and learning other mathematics topics (beyond algebra itself), and in accessing advanced mathematics topics, such as abstract algebra and linear algebra.

DOI: 10.4324/9781003197867-7

Subsequently, the role of teaching transformational algebra in development of ASS is analyzed, taking Kieran's perspective on the dual character of manipulative algebra. And finally in the last section, and based on the points made in the previous sections, we address the topic of teaching algebraic syntax. Here we take into account different approaches, some of which incorporate the use of technology-learning environments.

Transformative Algebra and the Presence of Structural Aspects in the Curriculum

This section shows examples of the presence of transformative algebra and the notion of symbolic algebra structure in some curricular proposals. Official and public national documents that express the basic curricular elements were reviewed, focusing the analysis on the paths of algebraic topics. It should be made clear that in none of the cases was an analysis of the enacted curriculum (Remillard & Heck, 2014) performed.

Curricular documents from four different countries were reviewed to show a plural outlook, that is: (1) American high school standards[1]; (2) the National Curriculum in England[2] particularly key stage 4, which shows the contents for grades 10 and 11 that correspond to ages 14 to 16; (3) the seventh national South Korean curriculum,[3] focusing the analysis on high school – grades 10, 11, and 12, ages 17 to 20 – and its six mathematics-related elective subjects that students can choose from; and (4) the Mexican curriculum and the reference courses of study of the Common Curricular Framework of Tertiary Education, developed by the Ministry of Education in 2017 (Secretaría de Educación Pública, 2017).

In reviewing the way in which algebraic content is organized in these curricula, we used the categories proposed by Bednarz, Kieran, and Lee, (1996) of the various approaches to teaching algebra, namely generalization, problem solving, modelling, and functional perspective.

The USA National Standards

Specifically, the CCSS.MATH.CONTENT.HSA.SSE. A.1, CCSS.MATH.CONTENT.HSA.SSE. A.2, CCSS.MATH.CONTENT. HSA.SSE. B.3, and CCSS.MATH.CONTENT.HSA.SSE. B.4 standards were reviewed. In the high school section, "Seeing structure in Expressions" is explicitly referred to as a specific topic of that educational level. In that regard, four topics are included on analysis of the structure of expressions and formulating equivalent expressions: (1) "Interpret expressions that represent a quantity in terms of its context" (CCSS.MATH.CONTENT.HSA.SSE.A.1); (2) "Use the structure of an expression to identify ways to rewrite it" (CCSS.MATH.CONTENT. HSA.SSE.A.2); (3) "Choose and produce an equivalent form of an expression to reveal and explain properties of the quantity represented by the expression" (CCSS.MATH.CONTENT.HSA.SSE.B.3); (4) "Derive the formula for the sum of a finite geometric series (when the common ratio is not 1), and use the formula to solve problems" (CCSS.MATH. CONTENT.HSA.SSE.B.4).

This curricular proposal in turn includes a set of transformative algebra elements, such as Arithmetic with Polynomials & Rational Expressions, Creating Equations, Reasoning with Equations & Inequalities, Operations & Algebraic Thinking, and Expressions & Equations.

Here, one should note the underlying hypothesis on the relationship between structural analysis and transformative algebra, which becomes patent when recognizing the structure of some expressions is requested, as can be seen in the following example:

"see $x^4 - y^4$ as $(x^2)^2 - (y^2)^2$, thus recognizing it as a difference of squares that can be factored as $(x^2 - y^2)(x^2 + y^2)$". In this case, the structural analysis of the expression will make it possible to perform transformative algebra activities, such as the building of equivalent expressions. The implicit premise that the analysis of the structure of an expression precedes its transformation shows the importance attached to the structural aspects of symbolic algebra in the standards, and the need for students to develop a sense of structure in order to become fluent in manipulative algebra. Whereas in the same standards, the relationship between polynomial zeros and coefficients is soon addressed to construct an approximate graph of the function defined by the polynomial.

Following the analysis of the structure of algebraic expressions, the introduction to algebra by creating equations and solving problems is proposed, in particular it is stated "Understand solving equations as a process of reasoning and explaining the reasoning; solve equations and inequalities in one variable, and solve systems of equations".

The creation of equations from the description of the relationships among numbers is another antecedent to solving simple equations. The creation of equations (modelling) is followed by the study of equation resolution and the representation and resolution of *inequalities*. Entry into resolution of inequalities is through graphical representation (as can be seen in the standard: CCSS.MATH.CONTENT.HSA.REI. D.10, where this is proposed from the outset: "Graph solutions to a linear inequality in two variables as a half-plane (excluding the boundary in the case of a strict inequality) and graph the solution set to a system of linear inequalities in two variables as the intersection of the correspond-ing half-planes"). Subsequently, working on performance of operations and algebraic thinking is proposed, by way of using basic operations with equations.

Although the attention paid to structural aspects of manipulative algebra, because of the number of issues related to problem solving, it can be said that this is the prevailing perspective in teaching symbolic algebra in the USA standards.

The UK National Curriculum

This curriculum is presented in a synthetic, non-prescriptive form and includes a comprehensive work proposal on missing value problems as of grade one of basic education. Solving equations with algebraic language starts from grade six and subsequently there is an emphasis on transformative algebra and problem solving. This curriculum does not include an analysis of the structure of algebraic expressions, and the study of algebra begins with the recovery of missing value problems to later start using and interpreting algebraic notation and understanding syntax in activities such as "writing $3y$ instead of $y + y + y$" or "writing a^2 instead of $a \times a$". Reference is made to the analysis of parentheses, but only as a sub-topic within the interpretation of syntax. After the introduction to the rules of algebraic notation, an emphasis is observed on problem solving. Rarely is the use of graphs proposed to understand the notion of "variable", and geometric elements are used to support the expression and solution of systems of equations.

In this curriculum, work with structural aspects is present through the solution of transformation exercises, albeit this feature is not made explicit.

Seventh South Korean National Curriculum

This curriculum proposal begins to solve simple equations as of grade two, although the study of algebraic language only begins as of grade six. The contents synthetically show an emphasis on modelling and real-life problem solving. The algebraic work is divided into the following fields: generalization, abstraction, analytic thinking, dynamic thinking, modelling, and organization (Lew, 2004). On the sequencing of transformative algebra, its study begins in grade 11, where reference is made to understanding the properties of powers, radicals, and the characteristics of systems of equations, matrices, and algorithms. In addition, it introduces problem solving that involves finding the general term in arithmetic or geometric sequences. In later grades, emphasis is placed on solution of problems involving equations with fractions and irrational numbers, and simple cubic equations.

This curricular proposal does not make work with structural aspects of symbolic algebra explicitly.

Common Curricular Framework for Tertiary Education in Mexico

The study of algebra in this curricular framework is introduced by way of characterizing phenomena of constant variation, to subsequently "symbolize and generalize linear phenomena and quadratic phenomena through the use of variables" (Secretaría de Educación Pública, 2017). Based on different modelling examples, the curriculum moves to algebraic thinking. The functional approach is adopted in the study of transformative algebra, an approach that is observed in sequences and numeric series themes (triangular, square, arithmetic, and geometric series) and their corresponding algebraic description. The algebraic treatment of word statements is a central and recurring theme in the curriculum of the Technological Baccalaureate (a sub-variant of the tertiary school system), revealing a hypothesis of justification for the presence of algebra in the curriculum as a tool for modelling and functionality. Whereas the secondary school curriculum (pre-tertiary school level) emphasizes solution of problems with everyday life problems, and with it confirming continuity from one school level to another in the problem-solving approach and functional approach to algebra.

Both the common curricular framework (tertiary education) and the secondary curriculum do not make structural aspects of symbolic algebra explicit and, in particular, no reference is made to the structural analysis of algebraic expressions.

As can be seen from the brief look at the cases reviewed, neither the structural aspects of symbolic algebra nor ASS – and even less so in this latter case – have become part of the mathematics curriculum. The USA standards are an exception to this as even if the term structure sense is not used, as of high school reference begins to be made to related activities, such as "seeing structure in expressions". That is, in these National Standards documents the role assigned to structural analysis of transformative algebra objects can be appreciated. Whereas in the other three curricula, implicitly the structural aspects underlie transformative algebra topics, which are subordinate to the dominant-specific approach in each case (i.e., problem-solving approach in the UK and South Korea, and modelling and functional approach in Mexico).

As explained in previous chapters, ASS includes a series of skills involving the recognition of structures in algebraic expressions (Hoch & Dreyfus, 2005), recognition in which algebraic substitution plays a decisive role. Typically, such perception and recognition skills are developed through intensive practice carried out at the teacher's request or at the student's initiative and almost never as an activity marked as compulsory in the curriculum.

A scenario such as the one seen in the revised documents suggests the need to bridge the gap between the curriculum and research results on structural aspects in teaching algebraic syntax. In Chapter 1 (this volume), Rojano has already raised the importance of thematizing ASS in the field of teaching, such as based on research, generalization, mathematics of variation, and modelling were thematized as well. However, unlike these "new topics" structure sense is not a matter of content, it is rather a set of capabilities, which represents the challenge of finding ways to make space in teaching for their development.

One step forward with regard to the aforementioned challenge has to do with one of the central ideas of this book, which is to deal with development of structure sense skills as an object of teaching. In this regard, Chapters 4 and 5 show how technology can play a relevant role, with a discussion on a variety of digital programmes and tools that can be used to promote organized transformative algebra practices leading to development of structure sense.

One must nonetheless admit that any innovative educational proposal must face another challenge: that of teacher training and education which, in the case of incorporating development of skills to perceive structure into the teaching of algebra, would have to include adequate communication of the characterization of activities that promote such development. Similarly, it is also essential to direct this communication to designers of educational materials and developers of digital applications for teaching mathematics.

However, in a scenario in which teaching algebra de-emphasizes symbolic manipulation, there is a need for arguments and especially examples that demonstrate the role of developing structure sense in students' mathematical maturity and in their ability to access mathematical content beyond algebra and beyond secondary education. The following sections present examples of transformative algebra in differential and integral calculus and analytic geometry and discuss research results on the relationship between high school ASS and university ASS.

ASS and Transformational Algebra in Post-Secondary School Topics

In addition to the importance of the skills inherent in structure sense due to their direct impact on the fluidity of work in transformative algebra, these skills play a decisive role in the resolution of tasks in other areas, such as integral calculus and analytical geometry, belonging to post-secondary mathematics. This section presents examples of tasks in these areas and seeks to illustrate the implications of transformative algebra and the structural analysis of algebraic expressions, in its resolution.

In basic courses of differential and integral calculus, in order to obtain the integral of a function resulting from a change of variable, recognition of the structure of the functional expression is usually used so as to identify a new variable and obtain the integral function using that new expression. For example, to solve the following integral

$$\int x\left(2+x^2\right)^2 dx$$

the idea is to recognize that a variable can replace a section of the functional expression. In this case, section $2 + x^2$ can be the new variable

$$v = 2 + x^2$$

whose derivative is

$$dv = 2x \cdot dx$$

Recognizing that for dv to appear in the function to be integrated, a factor of 2 is placed on it (and division by 2 is performed so as not to alter it) yielding

$$\frac{1}{2}\int 2x\left(2+x^2\right)^2 dx$$

Once the expression is complete, the integral is solved using variable substitution:

$$\frac{1}{2}\int v^2 \, dv = \frac{1}{2}\left(\frac{v^3}{3}\right) + c = \frac{v^3}{6} + c$$

Again, by replacing v with $2 + x^2$ the required integral is obtained:

$$\int x\left(2+x^2\right)^2 dx = \frac{\left(2+x^2\right)^3}{6} + c$$

This example illustrates how construction of the anti-derivative function is based on proper algebraic substitution, which depends on a structural analysis of the expression of the function to be integrated. One should note that here the transformative algebra activity focuses on building an equivalent expression through a change of variable that brings to light traits that enable it to be easily integrated, that is, incorporating features of visual salience. Thus, the action of integrating a function has components of ASS (algebraic substitution and visual salience) and the other calculation action – the derivation.

On the other hand, in basic analytic geometry courses it is important to recognize that an equation has a structure akin to a conical curve. For example, recognize if $9x^2 + 49y^2 + 18x + 249y - 9 = 0$ has the structure of an ellipse, such as:

$$\frac{\left(x+h\right)^2}{a^2} + \frac{\left(y+k\right)^2}{b^2} = 1$$

requires an analysis of algebraic expression $9x^2 + 49y^2 + 18x + 249y - 9$. In this case, one can see that the expression has numerical parts that are multiples and elements that can be grouped together until they reach the canonical form, as shown below:

$$9x^2 + 49y^2 + 18x + 294y - 9 = 0$$

$$9x^2 + 18x + 49y^2 + 294y - 9 = 0$$

$$9(x^2 + 2x) + 49(y^2 + 6y) - 9 = 0$$

In the next step, it's important to recognize that it is possible to add 441 to both sides of the equation in order to complete the squares:

$$9(x^2 + 2x + 1) + 49(y^2 - 6y + 9) = 441$$

$$9(x - 1)^2 + 49(y + 3)^2 = 441$$

$$\frac{(x - 1)^2}{49} + \frac{(y + 3)^2}{9} = 1$$

By recognizing from the onset that we can complete the expression to form two different squares, $(x - 1)^2$ and $(y + 3)^2$, it is possible to reach the result more expeditiously.

By its very nature, analytic geometry is a convergence of geometric and algebraic elements and, in the specific case of the previous example, one can see how analysis of the structure of an algebraic object (an algebraic expression) is guided by the search for a specific structure associated with a geometric object. As in the example of integral calculus here too transformative algebra leads to an equivalent equation, with features of visual salience that, in this case, make it clear that it is the equation of an ellipse.

Such integral calculus and analytical geometry exercises make it clear that development of ASS capabilities ceases to be an end in itself, as these capabilities are oriented by and combined with structures and concepts typical of other branches of mathematics.

On the other hand, as Hoch and Dreyfus (2004) point out, in the subject of algebra itself in post-secondary courses structure sense can be used in algebraic tasks that deal with solving equations of degree greater than 2, as it is shown by these authors in the following example.

In the equation $4x^2 - x^3 + 5(4 - 2x) = (3 - x^2)(6 + x)$, one can anticipate that, when removing the parentheses and performing the operations, the term in x^3 on the left side will be cancelled with the product of x^2 and x on the right side. With that transformation, second-degree equation $10x^2 - 13x + 2 = 0$ is obtained.

In another exercise, recognizing that the structure of fourth-degree equation $x^4 - 13x^2 + 36 = 0$ allows the variable change to be made $x^2 = m$, thus obtaining second-degree equation $m^2 - 13m + 36 = 0$.

In the field of functions one is often asked to find the zeros of a function, which involves recognizing that some of the terms can be grouped together, thus obtaining a factored expression. For example, to find the zeros of the function

$$f(x) = x^4 - 4x^3 + 2x^2 - 4x + 1$$

one can note that the terms x^4, $+2x^2$, and $+1$ constitute a perfect square trinomial, which is transformed into $\left(x^2 + 1\right)^2$. On the other hand, the missing terms $4x^3$ and $4x$ have as a common factor $4x$, and $4x\left(x^2 + 1\right)$ is obtained. By having this structure identified, the solutions to $x^2 + 1 = 0$ and $x^2 - 4x + 1 = 0$ can be obtained separately, thus finding the zeros of the original function.

The cases presented show how the ability to perceive or analyze the structure of expressions can be used, *inter alia*, to: (1) transform an expression and solve problems; (2) analyze properties of an expression or function of which it is a part; (3) create equivalent forms that allow solving problems and equations. As previously mentioned, the type of actions akin to those described in 1–3 are combined with actions and concepts typical of other areas of mathematics to solve various tasks.

ASS in Advanced Mathematics Topics

The examples given in the previous section demonstrate the role that secondary ASS skills play in resolution of algebra tasks in post-secondary courses. On the basis of a study involving pre-service mathematics teachers, Novotná and Hoch (2008) argue that structure sense of secondary and high school algebra can be considered a prerequisite for good performance in university-level abstract algebra courses. In that same article, the authors take up the definition of high school algebra (HSA) structure sense coined by Hoch and adapt it to develop a definition of university algebra (UA) structure sense.

Objects in university level algebra or abstract algebra are concepts; in fact, they are structures in and of themselves (e.g., the structure of group, ring or field), the instantiations of which are made up of sets of elements, operations, and properties derived from axioms.

The definition of UA structure sense proposed by the researchers applies to the elements of the sets and the notion of binary operation, on the one hand, and to the properties of binary operations (pp. 95–97), on the other.

In the study conducted with pre-service mathematics teachers, Novotná and Hoch (2008) applied a questionnaire with two items that identify the presence of HSA structure sense and two items that make it possible to identify the presence of UA structure sense. The outcomes they report show that, in most cases, those who successfully resolved the UA items also successfully resolved the HSA items, while those who did not successfully resolve the HSA items failed to do so with UA items as well.

While the data obtained and the theoretical arguments set out in the article suggest (as expressed by the authors) that HSA structure sense can be considered both a prerequisite for and a way to move towards development of UA structure sense, it is important to note that UA items are tasks involving symbolic algebra expressions the properties and transformations of which obey the rules of HSA. Thus, for example in item 3 of the same questionnaire "Let $x \, \Delta \, y = \frac{9x^2-16y^2}{6x-8y}$, decide whether Δ is a binary operation on the set of real numbers, and if so, determine its properties" (p. 100), verifying that operation Δ is binary in the real numbers requires checking that if x and y are real numbers, expression $\frac{9x^2-16y^2}{6x-8y}$ is an element of the real numbers. Resolution of the task depends on HSA knowledge and structure sense; in other words, they are directly related to resolution of the UA task. However, there are abstract algebra tasks in which the components that define a structure are of a different nature than that of known numerical sets, and their operations are not defined in terms of symbolic algebra expressions, for example, the elements of the symmetric group of permutations of a set X with n elements are not numbers but n-tuples of ordered symbols. In such cases, it is not clear that there is a relationship between application of HSA structure sense and verification of structural properties of the set – operation duo. This does not preclude conjecturing that there may be an indirect relationship between both structure senses – that of HSA and that of UA but testing it will require further studies that will likely be of a correlational nature.

And then just as HSA structure sense is a fundamental antecedent for solving a variety of abstract algebra tasks, UA structure sense also plays a decisive role in student performance in other areas of university mathematics. In this regard, Oktaç (2016) undertakes a comprehensive review of the literature on Abstract Algebra Learning, and in one of the sections of the article she refers specifically to the role of structural thinking in linear algebra activities. Results obtained from studies in which the researcher has participated reveal that some students face difficulties in trying to decide when a set of vectors is linearly independent in a vector space whose operations are explicitly given; students tend to propose equations using common addition and multiplication operations (Aguilar & Oktaç, 2004). That is, they leave aside the definition and verification of structural properties of the given operations.

Although works such as those of Oktac and collaborators attest to the relationship between the structure sense of abstract algebra and performance in other subjects of undergraduate mathematics, the review carried out by Oktaç shows that there is very little research on the subject so far.

Transformative Algebra and ASS

In Chapter 1, Rojano (this volume) mentions the trend in mathematics curricula that de-emphasized the teaching of manipulative algebra in the 1990s, to attach greater importance to conceptual aspects and aspects related to solving word problems. In this regard, Kieran (2013) extensively discusses features of procedural activity that are left hidden under the false dichotomy between conceptual understanding and procedural skills. Specifically, Kieran (2013, p. 212) points out that algebra has traditionally been considered an area where symbolic manipulation procedures predominate over the conceptual. Her counter-arguments to this vision of algebra are based on the work of two authors (cited in Kieran 2013): on the one hand, she bases them on the position adopted by Jean-Batiste Lagrange, who states that during their development the procedures are conceptual in nature; and on the other hand, on M. Donald's proposal that, once automated, procedures are updated and extended from conceptual elements.

In the field of empirical work, studies carried out by the same researcher, where she provides evidence of the conceptual presence underlying manipulative algebra and numerical skills, end up confirming the artificial aspects of the aforementioned dichotomy. Whereas, as already mentioned in Chapter 1 (this volume), results of the study on polynomial equivalence and rational algebraic expressions – undertaken by Solares and Kieran (2013) with 10th graders – show the emergence of conceptual knowledge during task resolution with the use of CAS and paper and pencil algebra. The conceptual knowledge in question (i.e., equivalence) was revealed in this study in two ways – syntactic (when polynomial and rational expressions are viewed as rational fractions) and numerical (when polynomial and rational expressions are viewed as functions), which, according to the authors, are attributable to the articulation of task design with the use of CAS technology.

By the same token, the statements made in Chapters 4 and 5 (this volume) on the Expression Machine (MEx) show that the co-existence of the procedural and the conceptual also takes place in activities aimed at promoting development of ASS. In MEx, tasks organized in multilevel networks promote practices related to structural aspects of algebraic expressions, creating an environment in which there is a convergence of manipulative algebra and an implicit use of fundamental concepts, such as equivalence and algebraic substitution. In some cases, task design tests the ability to perceive (by way of visual salience marks) structural elements of the algebraic expressions involved, but in others (and for most individuals) discovering or unraveling the structure of expressions requires symbolic manipulation work using paper and pencil. In other words, in the search to see the unseen, transformative algebra, which in its broad sense deals with syntax exercises involving an implicit use of concepts, plays an essential role.

However, based on analyzing the design of the three types of MEx tasks and the experience itself with this web environment, modalities of transformative algebra activity have been identified that relate to development of ASS (Chapter 5, this volume) and that are not based only on algebraic substitution or equivalence of expressions. These modalities consist mainly of comparing structures of given algebraic expressions, which are not connected necessary or only through relations of equality or equivalence, but by processes involving elements of reversible and hypothetical thinking. This is the case with MEx type II tasks, in which the generating expression and the output are given and one is asked to find the input expression(s). In them, the comparison of structures consists of recognizing structural features in the output that derive from applying the generating expression and thus formulating possible input expressions, which are tested, applying the generating expression, and corroborating that the output coincides with the given output.

Although MEx type II tasks can also be solved directly, solving for one of the variables the generating expression = output expression equation, there are subjects who use the strategy of developing a hypothetical expression and testing it. In turn, the design of type III tasks – input and output are given and the generating function must be found – also seeks to promote the comparison of structures and the formulation of hypothetical expressions, or to take advantage of visual salience features when any exist in the given expressions.

In a set of MEx type II and III tasks, intensive work with transformative paper and pencil algebra becomes necessary and, in such cases, transformative algebra not only conforms with conceptual aspects, but it also and most importantly does so with structural aspects.

ASS in Teaching Algebraic Syntax

This chapter on how HSA structure sense contributes to performing well in transformative algebra itself and in other mathematics and university algebra subjects highlights the importance of promoting development of structure sense capabilities throughout students' school life. It is particularly interesting to ask ourselves to what extent the type of initiation in the study of symbolic algebra influences further development of such capabilities.

In the early 1990s, a scenario in which the teaching of symbolic algebra was admitted to have failed and the findings from multiple studies evidenced the difficulties pupils faced when studying the subject, a group of specialists in algebraic thinking developed a research agenda based on a categorization of approaches to teaching and learning symbolic algebra. The members of this group conceived of approaches comprising activities of generalization, problem solving, modelling, and variation (functional approach) (Bednarz et al., 1996), each of which sought, among other things, to provide students with contexts in which they could make sense of algebraic symbology and its operations. The formulation of those approaches and their theorization can today be interpreted as a response to the critiques that prevailed at the time in respect of teaching based on rules of operation on chains of symbols that were devoid of concrete referents or of referents that were familiar to students.

A decade later, at the 12th ICMI Study meeting "The future of the teaching and learning of algebra" (Stacey et al, 2004), the *Approaches to algebra* working group questioned the previous classification, stating that those categories failed to capture a variety of approaches present in different countries' curricula, such as the French curriculum, in which equations are studied in the context of a formal approach to functions and their transformations, on the one hand, and that they failed to consider aspects of context, such as communities of students, teachers, and teacher trainers for which such classification makes sense, on the other (The Working Group of Approaches to Algebra, 2004, p. 7). As a result of work within the framework of the 12th ICMI study, in order to analyze approaches to algebra the group developed a toolkit that includes the diversity of participants' perspectives on school algebra and learning and teaching theories. The proposal emerged from the presentation and analysis of specific examples grounded in class practices, thus showing the value of recognizing the diversity of narratives.

Another topic related to algebra approaches was also addressed in the 12th ICMI study by the *Early algebra* working group, and it focused on analyzing the possibility of getting young students started in algebraic thinking (Lins & Kaput, 2004). Since then and throughout evolution of the research stream of the same name (early algebra), whether or not to include introduction to algebraic notation in that early initiation has been the subject of debate. In connection with that contention, in the *Early algebraization* field review book (Schmittau, 2011) reports results of a study based on implementation of Davydov's curriculum, which confirm Davydov's own findings in the sense that it is feasible to enrol elementary students in quantity comparison activities (before introducing them to the notion of number), in which they necessarily must use letters and not specific numbers to denote quantities and relationships between them. In other words, the data provided by such research support positions in favor of early algebra initiation, not only in theoretical or conceptual aspects, but also in its symbolic expression.

Although the above-mentioned approaches (widely disseminated and discussed) offer a wealth of sources of meaning for teaching algebraic symbolism, much less abundant is the literature on approaches to teaching algebraic syntax itself, which includes the algebra sign system, the chains of characters that make sense in that system and its transformation rules. A number of authors have used specific approaches in their theoretical or experimental studies, such as in the study on operations with unknowns involving students aged 12 to 13 where Filloy and Rojano (1989) used a version of the scale model and a geometric model for manipulation of two-figure parts with equal areas. In the first, using the equilibrium metaphor the actions of removing and putting weights on the scales recreate the Eulerian model of solving linear equations consisting of "doing the same on both sides"; while in the second, the actions of manipulating parts of the figures, maintaining area equivalence, recreate the Vietic model of solving equations consisting of transposition of terms from one side to the other of the equality sign. In both cases, it was possible to prove that the non-paradigmatic limitation of concrete models (i.e., their use does not apply to all modalities of equations, such as those with negative solutions or coefficients) results in difficulties for students to move towards the domain of syntactic models for solving linear equations with one unknown.

In the same sense as Filloy and Rojano's work with concrete models, the results reported by J. Vlassis (2002) in a study with a scale model confirm that students face difficulties in going from the concrete model to working with equations in symbolic algebra, primarily in cases of negative solutions. However, in her longitudinal study the author demonstrates that with adequate teaching intervention, students can overcome those difficulties. Similarly, the study carried out by Rojano and Martínez (2009) with a virtual version of the scale shows that with a guide provided within the same virtual environment it is possible for students to move towards a more abstract and widespread use of the model.

Other research provides data related to the difficulties faced by students in tasks involving grouping similar terms (Linchevski & Herscovics, 1996) or in the transition from the notion of equality in arithmetic to the notion of equivalence in algebra (Kieran, 1981). With a different perspective, Kirshner's study *The Visual Syntax of Algebra* tabled the topic of perception of structural aspects of algebraic symbology, reporting that while some students depend on the presence of visual salience traits to make decisions in their performance with algebraic syntax, others resort to solid propositional rules (Kirshner, 1989). These outcomes led the latter author to propose the dichotomy of a propositional introduction to the study of algebraic syntax, based on learning declarative rules, or an introduction in visual modality, as of training based on visual features that make structural aspects evident. Faced with this distinction, one would be tempted to think that the latter approach may be a better option to promote structural thinking as of basic education. However, Kirshner himself proposes to explore the transition between these modalities, thus implying that experience with both is necessary.

With regard to the possibility of facilitating the transition between different modes of learning and practicing algebraic syntax, Chapter 4 (this volume) analyzes a variety of learning environment designs using current digital tools that promote development of ASS with the support of elements of visual salience and, at the same time, fostering the learning, refinement and updating of transformative algebra techniques (in the sense of Lagrange, 2003) that come from learning rules and syntactic properties. Such is the case with the adaptive and tiered didactic design of the MEx and the design of algebraic sequences and tasks – with and without technology- using tools such as Aplusix (https://aplusix.org/), GraspableMath (https://graspablemath.com/) and platforms the likes of Kan Academy (https://es.khanacademy.org/).

Without underestimating the possible influence that a particular type of initiation in the study of syntax could wield in developing ASS among students, the truth is that technology learning environments now offer a wide variety of options for developing fluency in transformative algebra, favoring development of the ability to notice structural aspects of algebraic objects, and not only in the early stages but also at later stages of students' schooling. In this regard with respect to MEx, Muñoz and Xolocotzin (this volume) argue that "[...] technology can provide an organized, progressive, dynamic and individualized transition between the different paths as students resolve the tasks" and that "[...] In the future, this process may be carried out by using appropriate statistics and artificial intelligence".

In the same field of digital environments, one must also recognize the benefits of tasks in which structural features are worked with both in figurative, numerical and symbolic-algebraic contexts, as is the case of eXpresser, which is widely described and analyzed in Chapter 6 (this volume).

Notes

1 http://www.corestandards.org/
2 https://www.gov.uk/government/publications/national-curriculum-in-england-mathematics-programmes-of-study/national-curriculum-in-england-mathematics-programmes-of-study
3 http://ncic.re.kr/english.kri.org.inventoryList.do

References

Aguilar, P., & Oktaç, A. (2004). Generación del conflicto cognitivo a través de una actividad de criptografía que involucra operaciones binarias. *Revista Latinoamericana de Investigación en Matemática Educativa*, 7(2), 117–144.

Bednarz, N., Kieran, C., & Lee, L. (1996) Approaches to algebra: Perspectives for research and teaching. In N. Bednarz, C. Kieran, & L. Lee (Eds), *Approaches to algebra. Mathematics education library* (pp. 3–12), vol. *18*. Dordrecht: Springer. https://doi.org/10.1007/978-94-009-1732-3_1

Filloy, E., & Rojano, T. (1989). Solving equations: The transition from arithmetic to algebra. *For the Learning of Mathematics, 9*(2), 19–25.

Hoch, M., & Dreyfus, T. (2004). Structure sense in high school algebra: The effect of brackets. In M. J. Hoines, & A. B. Fuglestad (Eds.), *Proceedings of the 28th conference of the international group for the psychology of mathematics education* (Vol. 3, pp. 49–56). Bergen, Norway: PME.

Hoch, M., & Dreyfus, T. (2005). *Structure Sense in High School Algebra: The Effect of Brackets*. International Group for the Psychology of Mathematics Education.

Kieran, C. (1981). Concepts associated with the equality symbol. *Educational Studies in Mathematics, 12*(3), 317–326.

Kieran, C. (2004). The core of algebra: Reflections on its main activities. In K. Stacey, H. Chick, &M. Kendal (Eds.), *The future of the teaching and learning algebra. The 12th ICMI study* (pp. 21–33). Boston/Dordrecht/NY/London: Kluwer AP.

Kieran, C. (2013). The false dichotomy in mathematics education between conceptual understanding and procedural skills: An example from algebra. In K. R. Leatham (Ed.), *Vital directions for mathematics education research* (p. 153). NY/London: Springer.

Kirshner, D. (1989). The visual syntax of algebra. *Journal for Research in Mathematics Education, 20*(3), 274–287. doi: https://doi.org/10.2307/749516.

Lagrange, J. B. (2003). Learning techniques and concepts using CAS: A practical and theoretical reflection. In J. T. Fey et al. (Eds.), *Computer algebra systems in secondary school mathematics education* (pp. 269–283). Reston, VA: National Council of Teachers of Mathematics.

Lew, H. C. (2004). Developing algebraic thinking in early grades: Case study of Korean elementary school mathematics. *The Mathematics Educator, 8*(1), 88–106.

Linchevski, L., & Herscovics, N. (1996). Crossing the cognitive gap between arithmetic and algebra: Operating on the unknown in the context of equations. *Educational Studies in Mathematics, 30*(1), 39–65.

Lins, R., & Kaput, J. (2004). The early development of algebraic reasoning: The current state of the field. In K. Stacey, H. Chick, &M. Kendal (Eds.), *The future of the teaching and learning algebra. The 12th ICMI study* (pp. 47–70). Boston/ Dordrecht/NY/London: Kluwer AP.

Novotná, J., & Hoch, M. (2008). How structure sense for algebraic expressions or equations is related to structure sense for abstract algebra. *Mathematics Education Research Journal, 20*(2), 93–104. doi: 10.1007/BF03217479.

Oktaç, A. (2016). Abstract algebra learning: Mental structures, definitions, examples, proofs and structure sense. *Annales de Didactique et de Sciences Cognitives, 21*, 297–316.

Remillard, J. T., & Heck, D. J. (2014). Conceptualizing the curriculum enactment process in mathematics education. *ZDM, 46*(5), 705–718.

Rojano, T., & Martínez, M. (2009). From concrete modeling to algebraic syntax: Learning to solve linear equations with a virtual balance. In S. Swars, D. Stinson, & S. Lemons-Smith(Eds.), *Proceedings of the 31st conference of the North American chapter of the international group for the psychology of mathematics education*. Atlanta, Georgia, USA. 23 al 26 de septiembre. 5, 235–243.

Schmittau, J. (2011). The role of theoretical analysis in developing algebraic thinking: A vygotskian perspective. In J. Cai, & E. Knuth (Eds.), *Early algebraization. A global dialogue from multiple perspectives* (pp. 71–85). Heidelberg/Dordrecht/London/NY: Springer.

Secretaría de Educación Pública (2017). Planes de estudio de referencia del Marco Curricular Común de la Educación Media Superior.

Solares, A., & Kieran, C. (2013). Articulating syntactic and numeric perspectives on equivalence: The case of rational expressions. *Educational Studies in Mathematics*, *84*, 115–148. https://doi.org/10.1007/s10649-013-9473-7

Stacey, K., Chick, H, & Kendal, M. (Eds.) (2004). *The Future of the Teaching and Learning of Algebra. The 12th ICMI study*. New York, London, Kluwer A.P.

The Working Group of Approaches to Algebra (led by R. Sutherland) (2004). A toolkit for analysing approaches to algebra. In K. Stacey, H. Chick, & M. Kendal (Eds.) *The Future of the Teaching and Learning of Algebra. The 12th ICMI study*, pp. 73–96, New York, London, Kluwer A.P.

Vlassis, J. (2002). The balance model: Hindrance or support for the solving of linear equations with one unknown. *Educational Studies in Mathematics*, *49*, 341–359.

8 On the Connection of Structure Sense in Mathematics and Who Sees and Transforms It

Internal Structure, Visual Salience, and Transformative Algebra

Teresa Rojano

Introduction

This recapitulation has the purpose to recover ideas from the perspectives adopted in the different chapters and incorporate into the notion of structure sense elements related both to aspects of the intrinsic structure of mathematical objects (specifically objects of symbolic algebra) and to processes of semiosis and manipulation of objects, which occur when subjects interact with the latter. On the internal order, aspects of the grammar and the superficial and deep structures (parsing trees) of algebraic expressions are taken into account, as are the mathematical properties of equivalence, congruence and the action of algebraic substitution. In terms of the relationship of the subjects with the algebraic strings of characters, perceptual aspects such as the effects of visual salience, the linguistic experience of the subjects with the algebra sign system, and their ability to transform expressions in order to unravel, perceive and use their structural properties are considered. Additionally, the role played by technology learning environments in this subject-structure relationship is discussed, and the idea of conceiving "levels of structure sense" (which correspond to levels of complexity of the expressions themselves and algebraic tasks associated with them) is advanced. Furthermore, it is hypothesized that development of the algebra structure sense associated with these levels is feasible with the support of adaptive systems that consider, in each case, the structural features of algebraic objects and individuals' level of awareness of them.

Intrinsic Structure of Mathematical Objects: The Case of Symbolic Algebra

The word "structure" means *arrangement of and relations between parts or elements of something complex*. In this regard, the structure is endemic to mathematics, as is its abstract nature. In the case of algebra, the history of the (non-linear) evolution of this sign system, as Puig

DOI: 10.4324/9781003197867-8

describes in Chapter 2 (this volume), tells how these two characteristic features –structure and abstract nature – go hand in hand. According to Freudenthal's phenomenological perspective (1983), the concepts arise from the need to organize phenomena. In this sense, according to Puig's narrative, the structural properties of abstract algebra objects and their operations result from the objectification of modes in which other phenomena are organized through processes of abstraction. Puig provides an example with the concept of group, which was developed to study resolution and solvability of polynomial equations by radicals. In the beginning groups were a tool, before becoming objects as one of the structures of abstract algebra. The structure organizes a diversity of phenomena, namely all instantiations that possess the properties of being a non-empty closed set with respect to a binary operation that satisfies the associative property and fulfils the existence of a neutral element and a symmetrical element, hence its "more" abstract nature.

The above example shows that the emergence of abstract algebra concepts is closely related to the history of the evolution of the symbolic algebra sign system and, in particular, to the history of this sign system associated with the resolution of polynomial equations. As noted in Chapter 1 (this volume), this long stretch of history moved (though not linearly) from the expression of problems and their resolution entirely in words to the combined expression of words and symbols and, from there, to the fully symbolic expression of equations and their resolution. In that transition, the *texts* of the sign system moved away from their original referents; that is, from the referents of the contexts of money, trade, or debt problems (rhetorical algebra) and geometric referents (pre-vietic algebra).

It is this form of expression – symbolic expression – (far from its source referents) that has been integrated into the contents of school algebra and that is governed by the rules of syntax to generate expressions ("valid" strings of characters) and by transformation rules to manipulate such expressions. The difficulties that students face in learning and using such rules have been widely reported in the research literature, which has led some studies to focus on analyzing the relationship between subjects and the purely syntactic aspects of algebraic thinking. Such is the case with the studies of D. Kirshner, discussed extensively in the introductory chapter, which show that it does not suffice for a correct interpretation of algebraic expressions to master syntactic rules. Rather, there are features of deep structure (the parsing tree) and morphological structure (such as character spacing, parentheses or exponentiation) that strongly influence their reading by subjects (see, for instance, the article "The visual syntax of algebra", Kirshner, 1989).

However, in the interaction of subjects with the objects of symbolic algebra (expressions and equations), it is not only reading and interpretation process that are triggered, but transformative algebra actions are also taking place[1]. In this regard, another lesson from history is how in a general or algorithmic treatment plan of equation solving al-Khwārizmī (9th century algebraist) defines six canonical forms in his foundational algebra book on *Calculation by al-jabr and al-muqābala*, (Rashed, ed., 2007, p. 94 of the French translation), each of which is associated with a specific resolution method[2]. In today's terms, the foregoing translates into the strategy of performing equation transformations (for example, quadratic equations) that reveal versions that synthesize their internal structure (for example, the form $ax^2 + bx + c = 0$ or a factored version such as $(x - a)(x - b) = 0$) and which allow for application of a direct and specific method of resolution. That is to say, in addition to performing "correct" readings of the expressions involved, being proficient in solving equations means performing transformative algebra actions.

It is important to note that transformative algebra actions are usually performed in contexts in which they are aimed at a particular purpose. As such, the usual prompts in school algebra tasks are, for example, "simplify the following expression", "solve the equation below", "translate the next statement into an equation", each of which determines the type of transformative (or generative) action that leads to resolving the task. In this regard, transformative activity is not only about "blind and aimless" application of syntactic rules, but rather based on its guiding purpose the activity is located on a strategic plane, where – interestingly – the structure sense plays an essential role. Solares and Rojano (Chapter 5, this volume) describe the characteristics of algebraic tasks proposed in the MEx (mexalgebra.xyz) web environment, in which transformative algebra actions that lead to their resolution are guided by the need to compare structures corresponding to different expressions that are not necessarily equivalent. Some of these tasks consist of conceiving an algebraic expression as the output of applying a generating function to another expression (see section "Development of ASS, Levelling-Up Design and Task Types" of Chapter 5), which requires a high level of structure sense development that goes beyond the search for an equivalent expression or performing an algebraic substitution.

Similarly, the tasks of differential and integral calculus, analytical geometry, and linear algebra that are described in Chapter 7 (this volume) are examples of tasks in which it is necessary to carry out transformations that reveal very specific versions of the internal structure of an expression. An example that illustrates the foregoing tasks to recognize if $9x^2 + 49y^2 + 18x + 249y - 9 = 0$ has the structure of an ellipse, such as $\frac{(x+h)^2}{a^2} + \frac{(y+k)^2}{b^2} = 1$. As explained in the section "ASS and Transformational Algebra in Post-Secondary School topics" of the aforementioned chapter, in this case the structure sense consists of first noticing that the original expression has numerical parts that are multiples and elements that can be grouped together, with which transformations can be performed in the canonical form $9x^2 + 49y^2 + 18x + 294y - 9 = 0$, which can be transformed into $9\left(x^2+2x\right)+49\left(y^2+6y\right)-9=0$ and in which adding 441 to both sides of the equation to complete the squares, the equation $9(x-1)^2+49(y+3)^2=441$ is obtained, which can easily be transformed into the ellipse equation $\frac{(x-1)^2}{49}+\frac{(y+3)^2}{9}=1$. That chain of transformations of the original equation is guided by structural features of another algebraic equation whose referent is not an algebraic but rather a geometric object.

In the above example and in others described in Chapter 7, the presence of contextual factors guides and determines the type of interaction of the subject with the objects of symbolic algebra, an interaction involving structures expressed in the algebra sign system, but whose semantics pertain to another branch of mathematics. This gives rise to speaking about the "situated" character of structure sense, given that it comes into play in order to carry out transformations not in search of a single internal structure of algebraic expressions, but of a particular structure that has meaning in a particular context. This conception of different versions of the internal order of an algebraic expression contrasts with conceptions that are based only on the deep structure (parsing tree) of the expression or with the idea that transforming its surface structure leads to unraveling a single intrinsic structure.

In addition to the contexts mentioned so far – analytical geometry, differential calculus, linear algebra, and MEx tasks – there is resolution of word problems. To speak about the connection between the structure of relations that encompass the elements of a word problem and the structure of equations and algebraic expressions, Puig retrieves, from history, the problem solving–related projects of algebraists from different eras. As such, the latter author refers, on the one hand, to al-Khwārizmī's project as "[...] algebra is what people need to solve problems in which one has to calculate with amounts (of money, in all the examples given by al-Khwārizmī)" and, on the other, he cites a text from the work *Introduction to Analytical Art* by F. Viéte, in which this 16th century algebraist (considered the founder of symbolic algebra) makes it explicit that the aim of algebra is "the problem of problems, which is: To leave no problem unsolved" (Vieta, 1591, fo. 9r).

Puig points out that the two works referred to (seven centuries apart) recognize problem solving as the main purpose of algebra and that the problems that algebra deals with are those in which the value of unknown quantities is sought based on known quantities. With the evolution of the algebra sign system from its rhetorical to its symbolic version, problem solving now implies translating the expression of the problem in words into the expression in symbolic algebra code for the relationships among quantities – known and unknown – that are present in its statement; that is, in the form of an equation (or system of equations). This translation, involving a process of analysis[3] (Puig & Cerdán, 1988, 1990; Rojano

& Sutherland, 2001), ends when a quantity is expressed in two different ways that can be equaled (Puig, 1996). The process consists, among other things, of separating what is relevant from what is not relevant to the problem's solution (Puig & Rojano, 2004) and that is how the equation synthesizes the (internal) structure of the problem.

From its expression in equation form, the resolution is carried out entirely in the algebra sign system, without links to the referents of the statement's quantities. That process, which corresponds to solving the equation, is a process of synthesis[4]. Hence, it can be said that in this context, the interaction of subjects with algebraic expressions or equations takes place along two paths: one, where the algebraic code is used to express "the structure" of a problem (generative algebra activity); and another, in which the value of the unknown quantity is sought by way of solving the equation, applying the rules of transformative algebra. With exception of Chapter 6, so far, this book has dealt with the subject of structure sense in algebra, primarily in the field of reading and manipulating expressions and equations. Now the time has come to recognize that a structure sense treatment in the field of generative algebra (formulation of expressions and equations) is a topic pending further research.

Signification and Manipulation of Algebraic Objects

In order to broaden and deepen understanding of structure sense in algebra, the preceding section addressed the subject of interaction with objects of symbolic algebra from the standpoint of the structure and internal order of such objects. In this section, we analyze that interaction, taking into account the processes in which subjects build meaning related to the algebraic objects.

As explained in Chapter 5 (this volume), building meanings (or semiosis) with respect to algebraic objects can be understood from the adaptation of Pierce's semiotics to the case of algebraic language as a result of *sense* production, which takes place through acts of reading/transformation of the algebra *texts* (symbols, algebraic expressions, and equations). In this perspective, implementation of structure sense in transformative algebra actions is understood in terms of those processes of signification associated with structural aspects of the algebraic objects.

In the previous section, reference is made to the "situated" nature of structure sense, given that noticing certain structural features is guided by the context and the task watchwords, that is to say it is placed on a strategic plane. Yet, on the other hand, according to the aforementioned semiotic perspective, in addition to the structural attributes of algebraic objects and the contextual factors, the subjects' linguistic and mathematical experience is also involved in the processes of signification. That is, in the same task and context, the structure sense does not necessarily unfold uniformly in different subjects. The cases of Francisco (F) and Lulu (L) described in Chapter 5 illustrate this, showing well-differentiated strategies *vis-á-vis* the same set of MEx tasks, designed for users to bring into play and develop their structure sense.

With long experience as a primary school teacher with graduate studies in mathematics education, L successfully solved the tasks of an arithmetic nature. After several attempts, she manages to solve some of the items in the set of algebraic tasks that require comparison of structures accompanied by logical reasoning, investing considerable time and resorting to the direct (algorithmic) application of manipulative algebra rules and, in some cases, taking advantage of the visual salience traits of expressions. Whereas F, with training as a high school mathematics teacher with graduate studies in mathematics education, when faced with the same set of tasks showed a tendency to use strategies that involve capabilities of perception and/or analysis of the structure of the expressions involved, analysis based on the comparison of those structures with each other (see section "Francisco, Lulu and MEx Tasks", Chapter 5).

As explained in that very same Chapter 5, in terms of semiosis processes, F and L's approaches to the same MEx task group involve acts of reading/transforming algebraic texts (the MEx tasks and expressions), acts that are influenced by reference to other texts and their significations residing in each of their mathematics experiences. That is, F and L embark on resolution routes in which they are immersed in intertextual networks based on which processes of sense production and meaning construction take place with respect to the internal order of the expressions involved in the tasks proposed in MEx. In L's case, one can say that this intertextuality is permeated primarily by senses and meanings derived, on the one hand, from her experience with arithmetic objects and operations and, on the other, from her fluency with the (almost algorithmic) application of the syntax rules of symbolic algebra. Meanwhile in F's case, one can hypothesize that his prolonged interaction with manipulative algebra (due to his teaching activity) generated intertextual networks derived from "habits" of inspecting and analyzing algebraic expressions and equations.

Beyond recognizing that verification of the assumptions in the cases of F and L merits a treatment of the data that delves into the type of texts and significations evoked along the respective resolution paths, one can assert that with the above semiotic approach that includes the notion of intertextuality[5] it is possible to understand in transformative algebra contexts the ability of subjects to notice the internal structure of algebraic objects (structure sense) in relation to dynamic (and situated) processes of assigning meanings to those objects.

In this section, it's been suggested to address the connection of structure to the person seeing and transforming it – the subject matter of this chapter – with an alternative (but complementary) perspective to

constructivist or sociocultural views that have long dominated literature specialized in the interpretation of algebraic symbology. One of the
potentialities of such alternative vision resides in its specificity, because
the theory that supports it is a theory of signs, which in its adaptation
translates into a theory of algebraic signs (and texts).

Despite the relevance and potentialities of the oft-referred semiotic
approach, analysis of factors of a cognitive nature is also necessary to
broadly understand the subject-structure relationship. Kirshner makes
this point, emphasizing the importance of studying, for instance, visual
perception of algebraic expressions with elements of visual salience, such
as character spacing, exponentiation, or grouping by means of parentheses or other modes of writing. However, treatment of this type of
cognitive aspects is beyond the scope of this chapter.

Structure Sense in Other Contexts and Sign Systems

Structure Sense and Figurative Patterns

A previous section analyzed tasks showing the situated nature of structure
sense in algebra (ASS). In resolving these tasks, the ASS reveals itself as a
capability to notice the structural aspects of expressions and equations,
a capability that is guided by the context and purpose of the task and in
which the referents of algebraic texts come from other areas of mathematics. Examples of this are tasks that seek to prove that a given equation
represents an ellipse or that a function is the antiderivative of another
function. As a complement to the above, reference is made here to research
in which in the design of activities for experimental work symbolic algebra is related to other sign systems and in which referring to structure
sense entails algebraic objects and objects belonging to those other systems as well. Such is the case of the eXpresser microworld activities that
are described in detail in Chapter 6 (this volume) and in which numerical,
figurative, and symbolic representations co-exist dynamically.

The study carried out with eXpresser begins with the conception of
generalization as "the process of noticing the structure of a figural pattern, identifying what is repeated and expressing the rule that corresponds to this structure symbolically". Thus, the activities are designed
to bridge the gap between identifying and expressing a pattern. A feature of this design is that, given a figurative pattern that unfolds on the
computer screen, students are asked to reproduce it by creating their
own model, based on the way(s) in which they perceive its structure (see
Figure 6.1, Chapter 6). The microworld provides a variety of tools to
express the generality of that structure and with it an environment is
created in which students manage to connect the structure of the pattern
they perceive in the figurative version and the structure of the expression
(symbolic or numerical) of the rule that they use to describe it. Thus

just like in the case of solving word problems, in eXpresser activities the symbolic expression synthesizes the relationships among elements of an object in a context other than algebra.

Both in solving word problems and in eXpresser activities, algebra plays the role of a medium for expression and in the latter generative algebra actions go through what Filloy et al. (2008, p. 221) call "intermediate language strata" between a more concrete level sign system (in this case, that of the figurative pattern) and a more abstract level sign system (that of symbolic expression). In other words, production of the general rule does not result from a direct translation to the algebra sign system. Rather, facilitated by eXpresser's own technological and dynamic environment, students make use of tools that allow them to express the structural properties of the pattern in ways in which proximity is maintained with the referents in the figurative version albeit without directly dealing with a formal version of the algebraic syntax (Geraniou & Mavrikis, this volume). From a cognitive perspective, students' performance with generative algebra activities in eXpresser can be thought to correspond to what Rivera (2010) calls "*abductive-inductive action on objects*"[6] and it is a clear example of the close relationship structure–abstract nature of the mathematical activity.

In the same field of cognition, the findings of the aforementioned research show that students identify (i.e., perceive) different units of repetition in the same figurative pattern, giving rise to rules describing the pattern that reflect different versions of its structure. These data show that visual salience of structural aspects is not a unique intrinsic attribute of a figurative pattern; rather, it is something relative, depending on the person perceiving it. As the researchers of the study with eXpresser point out, findings the likes of this deserve to be analyzed from the standpoint of cognitive theories specifically related to perception phenomena.

On the other hand, in the field of conceptual mathematics, didactic design of the study's collaborative activities contemplates having students compare with classmates and justify among themselves their respective models and descriptive rules of a same pattern, thus creating a situation conducive to strengthening the notion of equivalence, which is fundamental to development of structure sense (be that in algebra or in other contexts). In this case, verifying the equivalence between the rules is not achieved by transforming one into the other (as would be done in algebra), but it is based on the fact that the rules, albeit each different from other, represent the same figurative pattern.

Structure Sense and Numerical Thinking

Although a body of research in the field of history and education has marked clear differences between arithmetic and algebra, dialectically these domains of mathematics are at the same time closely related to each other as forms of thought. In particular, numerical thinking is a context that naturally relates to ASS. The study reported by Kieran and Martínez in Chapter 3 (this volume) shows that by engaging in carefully designed numerical activities, sixth grade students develop an ability to notice the structural properties of numbers as entities and numerical operations, such as multiplication and division.

As for noticing the internal structure of numbers, this capability allowed the students participating in Kieran and Martínez's research to decompose and rewrite a number in multiple ways and thereby develop a sense of equivalence of expressions, which is a notion underlying the ASS. While their approach to the concept of the inverse operation of multiplication is part of a structure sense with respect to the sign system of arithmetic, in this case of the set of natural numbers and multiplication and division operations. One can hypothesize that these two types of structure sense will play a significant role in the development of ASS. However, this would require strict longitudinal research that follows up on the same generation of students with experience in developing their number structure sense in a pre-algebraic stage, and analyzing the recovery of that experience in their work with activities that promote development of the ASS.

Based on studies conducted in the 1980s and 1990s showing essential differences between arithmetic and algebraic thinking (Filloy et al., 2008; Sfard & Linchevski, 1994, among many others), one can anticipate that such differences will be manifested in development of the structure sense in algebra. Among these differences or changes in thinking are that of working with unknowns and that of algebraic symbology. In this regard, Radford (2011) reports that in the context of generalization activities with figurative patterns, students face dealing with what he calls indeterminacy (the condition of "unknown" of the indeterminate quantities in the sequence) and analyticity (the way to treat the unknown as if it were known). In other words, what is reported and argued theoretically by the latter author implies that empirical studies that address the problem of developing structure sense in the arithmetic-algebraic field will have to take into account continuities and discontinuities between one form of thinking and the other. In this regard, the next section discusses how technological environments and dynamic versions of concepts that can be recreated in those environments can facilitate the transition between these forms of thinking.

In another vein, but again in relation to the role of numerical sense in algebra, it is important to note that different sets of numbers (natural, integer, rational, irrational, real) are part of the sign system of symbolic algebra. Numbers appear in expressions and equations as coefficients, exponents, and roots (in the case of equations) and, consequently, the structure sense of numbers and their operations is imbricated with the ASS. The aforementioned task of verifying whether equation $9x^2 + 49y^2 + 18x + 249y - 9 = 0$ is equivalent to ellipse equation $\frac{(x+h)^2}{a^2} + \frac{(y+k)^2}{b^2} = 1$ illustrates this imbrication in the transformation steps of the initial equation, in which the perception of the internal structure of the numbers 9, 49, and 18, relating them as multiples of 9, is indispensable to obtain equivalent equation $9\left(x^2 + 2x\right) + 49\left(y^2 + 6y\right) - 9 = 0$, which with subsequent transformations leads to the equation of the ellipse.

In short, structure sense in numerical thinking is not only an important antecedent to developing ASS, it is also part of it.

Creation of Textual Spaces for Development of Structure Sense

The Role of Digital Technologies

Elements collected from previous chapters have allowed us to delve deeper into understanding the notion of ASS, based on the analysis of its relationship with performance in tasks of differential and integral calculus, analytical geometry, linear algebra, and MEx and eXpresser tasks (the latter, generalization tasks with figurative patterns). This analysis revealed the situated nature of ASS, as well as a number of capabilities related to this structure sense, such as the ability to compare structures of algebraic expressions with each other and with object structures (mathematical or not) from other contexts. In addition to revealing new aspects of ASS, this poses significant educational challenges to help promote development of this capacity. In this regard, it is worth recalling ideas expressed in Chapters 4, 5, and 6 concerning the potential of technological learning environments, such as the possibility of designing multilevel task networks, use of feedback tools, and adaptive and gamification systems (see Muñoz & Xolocotzin, and Solares & Rojano, this volume), as well as the possibility of designing environments in which different representations and mathematical objects structures and concepts are smoothly transitioned (see Geraniou & Mavrikis, this volume).

From the semiotic perspective described in other chapters, using the technological potential of digital programs and tools would mean creating learning environments conceived as *textual spaces* that support students (or users in general) in developing their ASS. Puig (1994) introduces the term *textual space* to the field of mathematics education. It comes from the idea used in modern literary theory that a *text* is open; it means that it is "opened" by its relations with other texts that the reader knows from his/her previous experience, quite to the contrary of the idea that the reader, through the act of reading, extracts an intrinsic meaning from the text (notion of closed text). That open text before the reader, related to other texts that the reader brings to mind is what is called textual space. The textual space has empirical existence and imposes a semantic constraint on those who read it; and the text (resulting from the reading/transformation of textual space) is a new articulation of that individual space in which other texts intervene. Based on this notion of open text, a mathematical task (which has a didactic intent) can be considered a textual space to be read and transformed into other texts imbued with sense (and meaning) by its intertextual relationship with other texts previously produced (Rojano et al., 2014).

In this perspective of intertextuality, each MEx task is a textual space composed of numerical and algebraic objects (expressions and equations) that are located in a multilevel network of tasks. The design seeks to have subjects bring into play their ability to perceive or discover the structure of the objects in question through their reading/transformation of the task. That is, each task is a space open to texts and their meanings, which have been produced previously. The actions of L and F in MEx tasks, described in Chapter 5, show a personal intertextual activity, in which each of them evokes meanings (and senses) constructed in their mathematical experience, which includes the resolution of previous tasks of the network.

As already mentioned, the paths of L and F through the MEx task network, as well as the strategies used to solve them, are essentially different from each other and in each case the path reflects part of the intertextual network in which each of them is immersed. This information is useful, on the one hand, in the field of technology for designing an adaptive system that can be integrated into the web environment; and, on the other, in the field of research to monitor and analyze how individual use and development of the ASS evolves.

As for a future agenda, the design of the multilevel network of tasks in MeX, based on levels of syntactic complexity of the algebraic expressions involved and the type of relationships among them, as well as the result of analyzing the cases of L and F all set the tone for further data analysis in large groups of users. A large-scale study would make it possible to categorize the actions of users and enable characterization and definition of ASS development levels. It would contribute to advancing research on the subject of ASS and the design of technology learning environments that promote its development.

Now with regard to extending the theoretical view of ASS in other contexts, one can assert that the eXpresser microworld is a textual space composed of mathematical objects of a numerical, figurative, and symbolic nature, in which students enrol themselves in acts of comparing the structures of these objects. These acts of comparison are acts of reading/transforming texts belonging to different sign systems and, here, intertextuality consists of reference to texts previously read and transformed into those sign systems. In particular, what is observed in the data reported from experience with eXpreser is that there are several "readings" of the same figurative pattern, which can be interpreted as being influenced by previous experiences of subjects with configurations of that type and/or by their direct and dynamic interaction with the elements of that configuration. It bears remembering that this dynamic interaction takes place during subject construction of the model or models that reproduce the given configuration. In other words, direct manipulation of the pieces that make up the figure can trigger processes of semiosis (production of meaning) that result in different versions of its structure. The feedback tools – verifying the validity of the model built (display of the structure found by coloring in the figure) – encourage students to compare different versions of the structure of one same pattern, both visually and by comparing the rules developed. In addition to producing meaning on structural aspects of the pattern, work in the eXpresser microworld allows students to advance their knowledge of the concept of equivalence, both of which are considered essential parts of structure sense.

MEx and eXpresser designs are examples of how digital environments can be used to foster development of structure sense in students, not just in reference to the algebra sign system, but also to the relationship of that sign system with structural aspects of other sign systems, be they of a mathematical nature or not. Experiences with these environments – reported in previous chapters – together with the analysis presented in Chapter 4 of the innovative features of the technological tools available offer a set of real possibilities for diving deeper into the field of research and didactic design, aimed at understanding and promoting the mathematical maturity of students by developing their ability to perceive and use the structure of mathematical objects.

To close this last chapter, it is worth referring back to the episode of Sonya L., the gifted girl from the Krutetskii study mentioned in the introductory chapter, in order to approach the question of whether it is possible for non-gifted subjects to develop the structure perception skills the likes of those shown by Sonya. The theoretical and empirical insights discussed throughout the eight chapters contribute elements and principles of didactic design in environments – be they technological or not – that support development of ASS among diverse learners, with which we hope to have provided a reasonable response to the question posed.

Notes

1 C. Kieran (2004) classifies algebraic activity as generative (related to the algebraic interpretation and representation of situations, properties, patterns, and relationships); transformative (relating to all types of symbolic manipulation); and global/meta-level (involving more general mathematical processes related to contexts than motivate the use of algebra).

2   In the first part of his foundational book *The Compendious Book on Calculations by* al-jabr *and* al-mukabal, al-Khwarizmi explains the solution of the following six types of equations that are the forms to which all linear and quadratic equations can be reduced (canonical forms): treasure equals roots, treasure equals numbers, roots equals numbers, treasure and roots equals numbers, treasure and numbers equals roots, and roots and numbers equals treasure. In modern notation, these types are: $ax^2 = bx$; $ax^2 = b$; $ax = b$; $ax^2 + bx = c$; $ax^2 + c = bx$; and $ax^2 = bx + c$ (van der Waerden, 1985).

3   The *analysis* process leads to a translation process of an arithmetical nature, which consists of transforming the initial text of a problem in a new text, in which the elements that intervene in more elementary translations are made explicit, in order to make explicit as well, the ways these elements are linked within the arithmetic expression that solves the problem (Puig & Cerdán, 1990, pp. 38–39).

4   The *synthesis* process is the inverse process of the *analysis* one and consists of performing the operations of the arithmetic expression (which involves only known quantities) derived from the process of *analysis* (Puig & Cerdán, 1990, pp. 38–39).

5   In his book *Intertextuality*, Graham Allen (2000) points out that *intertextuality* "[...] has been defined so variously, that it is anything but a transparent, commonly understood term" and explains how the term has been used in different theories; semiotics is one of those theories. In an adaptation of the semiotic interpretation of the term to the algebraic reading and writing, an *intertext* is the result of the act of reading/transformation of an initial text opened and related to other texts, whose sense production stems from the relations among the texts involved in that reading/transformation process (Rojano, Filloy, & Puig, 2014).

6   Rivera (2010) refers to an *abductive-inductive action on objects* as an action that involves "employing different ways of counting and structuring discrete objects or parts in a pattern in an algebraically useful manner" – an action that allows subjects to perceive the structure of a pattern, generalize it, and symbolize such generalization.

References

Allen, G. (2000). *Intertextuality*. London: Routledge.

Filloy, E., Rojano, T., & Puig, L. (2008). Educational algebra. *A theoretical and empirical approach*. New York: Springer.

Freudenthal, H. (1983). *Didactical phenomenology of mathematical structures*. Dordrecht: Reidel.

Kieran, C. (2004). The core of algebra: Reflections on its main activities. In K. Stacey, H. Chick, &M. Kendal (Eds.), *The future of the teaching and learning algebra. The 12th ICMI Study* (pp. 21–33). Boston/Dordercht/NY/London: Kluwer AP.

Kirshner, D. (1989). The visual syntax of algebra. *Journal for Research in Mathematics Education, 20*(3), 274–287.

Puig, L. & Rojano, T. (2004). The history of algebra in mathematics education. In *The Future of the Teaching and Learning of Algebra The 12thICMI Study* (pp. 187–223). Springer, Dordrecht.

Puig, L. (1994). *Semiótica y matemáticas [semiotics and mathematics]*. Valencia: Episteme.

Puig, L. (1996). *Elementos de resolución de problemas [Elements of Problem Solving]*. Granada, Spain: Comares.

Puig, L., & Cerdán, F. (1988). Problemas Aritméticos Escolares *[School arithmetic problems]*. Madrid: Ed. Síntesis.

Puig, L., & Cerdán, F. (1990). Acerca del carácter aritmético o algebraico de los problemas verbales. In E. Filloy, & T. Rojano (Eds.), *Proceedings of the II Simposio Internacional sobre Investigación en Educación Matemática* (pp. 35–48). PNFAPM – Universidad Autónoma de Morelos.

Radford, L. (2011). Grade 2 students' non-symbolic algebraic thinking. In J. Cai, & E. Knuth (Eds.), *Early algebraization* (pp. 303–322). Berlin: Springer-Verlag.

Rashed, R. (Ed.). (2007). Al-Khwārizmī. *Le commencement de l'algèbre*. Paris: Librairie Scientifique et Technique Albert Blanchard.

Rivera, F. (2010). Visual templates in pattern generalization activity. *Educational Studies in Mathematics*, 73(3), 297–328.

Rojano, T., Filloy, E., & Puig, L. (2014). Intertextuality and sense production in the learning of algebraic methods. *Educational Studies in Mathematics*, 87(3), 389–407. doi: https://doi.org/10.1007/s10649-014-9561-3.

Rojano, T., & Sutherland, R. (2001). Arithmetic world – Algebra world. In H. Chick, Stacey, K, Vincent, J & Vincent, J (Eds.). *The future of the teaching and learning of algebra*. Proceedings of the 12th ICMI Study Conference, vol. 2, pp. 515–522. University of Melbourne.

Sfard, A., & Linchevski, L. (1994). The gains and the pitfalls of reification – The case of algebra. *Educational Studies in Mathematics*, 26, 191–228.

van der Waerden, B. L. (1985). A history of algebra. *From al-Khwarizmi to Emmy Noether*. Heidelberg: Springer-Verlag.

Vieta, F. (1591). *In artem analyticem isagoge*. Turonis: Iametium Mettayer Typographum Regium.

Index

For Product Safety Concerns and Information please contact our EU
representative GPSR@taylorandfrancis.com
Taylor & Francis Verlag GmbH, Kaufingerstraße 24, 80331 München, Germany

* 9 7 8 1 0 3 2 0 5 5 1 1 4 *